COMPLETE GUIDE TO TRIUMPH HERALD & VITESSE

Mike Costigan

Bay View Books

Published 1992 by Bay View Books Ltd
13a Bridgeland Street
Bideford, Devon, EX39 2QE

© Copyright 1992 by Bay View Books Ltd

Designed by Peter Laws

Computer typesetting and layout by Chris Fayers

ISBN 1 870979 27 3
Printed in Hong Kong

Contents

Introduction	7
Using This Book	8
CHAPTER ONE The Models	10
CHAPTER TWO Colour Schemes	25
CHAPTER THREE Bodywork	27
CHAPTER FOUR Exterior Trim	39
CHAPTER FIVE Exterior Fittings	54
CHAPTER SIX Wheels	60
CHAPTER SEVEN Mechanical Components	63
CHAPTER EIGHT Interior Trim	78
CHAPTER NINE Dashboard & Controls	91
CHAPTER TEN Optional Extras	100

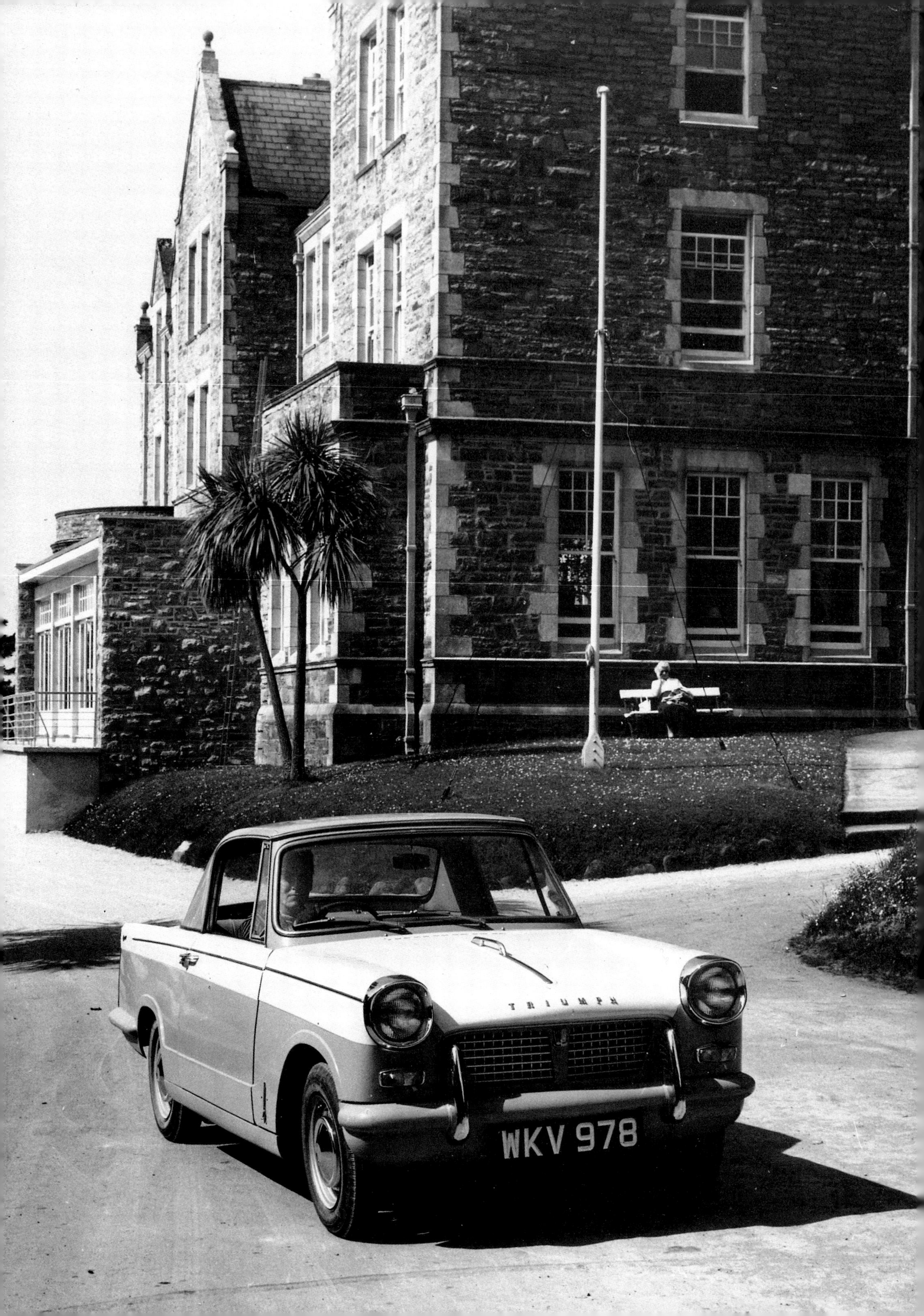
TRIUMPH
WKV 978

INTRODUCTION

My first memories of the Triumph Herald are of a Powder Blue chassis under the spotlights in the showrooms of a Grantham dealer. Presumably the new saloon and coupé were also on display, but, as a 12-year-old, my attention was drawn to the procession of gaily-coloured Dinky toys across the low windowsill. Perhaps these memories would have had no lasting quality, but my father was looking at new cars to replace our faithful 1936 Lagonda Rapier.

The Herald proved sufficiently attractive for him to place an order for our first new car, and eventually our Black twin-carb saloon with leather upholstery arrived, albeit in Coffee with Vynide trim! The build quality of these early cars left something to be desired, and our saloon was no better than most. According to my father's diary, within the first six months the exhaust was welded twice, the diff, steering wheel and driver's seat were all replaced, and the car was given a complete respray. However, once the teething troubles were sorted out, the car gave reliable service, and was to provide the foundation for the family's loyalty to the Triumph marque. The 948 saloon was replaced by one of the first 12/50s, on which I learned to drive, followed by a 2000 and a Dolomite; the 12/50 passed to my elder brother, who replaced it with an early Vitesse 1600, and now owns a Dolomite and a 1500.

My own early motoring was done in a succession of pre-war cars, mostly Austin Sevens, although I eventually succumbed to the charms, and power, of a 'modern' car, in the form of a six year-old SAH-tuned 2-litre Vitesse convertible. This was replaced by one of the first Dolomites, but I then deviated from the Triumph range until I was tempted by a forlorn Bond Equipe tucked behind a local garage. This was the catalyst that rekindled my enthusiasm for Triumphs, and since 1981 I have owned eleven Heralds, one Vitesse, one GT6 and three Spitfires.

From this list it can be seen that the Herald is my favourite model – and the coupé is my favourite Herald. Indeed, at one stage there were five coupés in the drive! This collection has since been rationalised to the present two, an early 1959 Dublin-assembled version and a mid-1960 version. These are accompanied by a Courier, together with a couple of near relations in the form of two Standard Atlases. If there was some doubt about my sanity, ownership of the latter must surely remove any question mark!

I have always found the history and detail changes in cars fascinating. Perhaps if this book had been written 15 years ago, the subject matter would have been the Austin Seven, but in recent times my enthusiasm has been channelled towards these Triumphs. Nevertheless, it could not have been written without a lot of help from others, and I would like to take this opportunity to thank some of them here.

First, I would like to acknowledge the tremendous influence that John Griffiths has had on the Triumph world – he was the leading light in the Triumph Sports Six Club for 14 years, and can shoulder much of the responsibility for the popularity of the Herald and Vitesse today. He also encouraged my own enthusiasm, and was largely responsible for my present position of Archivist within the TSSC. John Kipping is another enthusiast who has done much to preserve and maintain our cars on the road, and he has been good enough to read through the manuscript, correcting several errors, and adding more details of which I was ignorant. The format of the book has been influenced by work done within the TSSC by Chris Longhurst, the Herald Registrar, and by the former Vitesse Registrar, John Thomason.

I would also like to thank those many owners who have suffered my prying into their pride and joy, especially Peter Jevons, who provided useful additions to my library, and Chris Allen and Richard Lee, who were subjected to more aggravation than most! Anders Clausager has been most helpful in making the archives of BMIHT available, and provided much of the detail in Appendix III; the TSSC archives have also provided much of the historical detail. Last, a special thank you to my wife Isabel, who has survived the upheaval to our daily routine, and corrected and typed the manuscript.

USING THIS BOOK

This book has been produced to provide a general guide to originality of specification and presentation relating to the Triumph Herald and Vitesse. It should help those owners who are anxious to maintain or rebuild their cars to as close to factory specification as possible. Hopefully, it will provide the answers to most questions, but it will doubtless cause some owners to query my authority!

I must emphasise that the information contained here should be used as a guide only. Production line methods frequently resulted in vehicles being made with discontinued components, and during the 1950s and early 1960s especially the Triumph Works could usually be persuaded to produce a vehicle to bespoke specification – for instance, if the order came through an influential distributor.

By the very nature of their construction, the Herald and Vitesse models lend themselves to non-factory permutations, so that 'semi-official' Vitesse estates produced by the Standard-Triumph London Service Centre, for example, should certainly be considered to be 'original', even though they did not leave the Coventry factory in this form. A Vitesse coupé could have been produced just as easily, and indeed the official production statistics show just such a car in the 1963 figures. However, it is quite possible that this is the prototype Vitesse, which was built in 1959 on a left-hand drive 948 coupé, but not issued with an English registration until late 1962 or early 1963.

I would hesitate to condemn a vehicle as 'non-original' without a detailed inspection: if unrestored, then original features and fittings usually look right – and if it looks right, the chances are it probably is. So if you find a vehicle which appears to go against the specifications suggested in this book, share your find with other knowledgeable enthusiasts before you strip it down for that 'full rebuild'! Once restored, the evidence is destroyed and one can only judge the vehicle on its likely original specification. If you want to avoid arguments over doubtful features, this book will guide you in producing the authentic presentation expected in today's concours events.

Throughout the book, references to the different models will be simplified: 948, 1200, 12/50 and 13/60 will serve for the relevant Herald models, although the first was only known by the title 'Herald'; 1600, 2-litre and MkII will simplify reference to the Vitesse range, although the early car's correct title was 'Vitesse 6'. Where reference is made to different chassis designs, MkI and MkII (Roman numerals) will be used for Vitesse versions produced up to HC12079 and after HC50001, whereas Mk1 and Mk2 (Arabic numerals) will be used for Herald versions produced before and after GA80000. This should avoid the confusion and mistakes which occur with verbal references to the different chassis.

BELOW
The Vitesse estates converted at the Park Royal factory, and marketed through the Standard Triumph London Service Centre, featured carpeted panels between the rubbing strips, and these full-length rear wood cappings.

INTRODUCTION

LEFT
This delightful shot shows the Herald range of early 1960 – although to North American export specification, including white-wall tyres and white rubber bumpers. Key identification features include the early coupé roof style (lacking the horizontal ribbing introduced in June 1960), the two-tone upholstery visible in the convertible (unique to the 948cc models) and the bonnet handle without additional bright trim on the saloon. The colours, too, are interesting: Coffee was unique to 948s, Powder Blue continued into 1200 production, while Signal Red was the only colour to be available from the earliest 948 to the last Vitesse.

Where dates are used to identify production changes, these relate to factory production dates – registration dates follow these, sometimes by several months. Where possible commission numbers will also be quoted, but these relate to the first fitting of the relevant change; vehicles with earlier numbers would definitely not have the new feature, but subsequent numbers may well use the discontinued design. Occasionally a gap in number will appear before a major change, and in these cases later numbers will all feature the new specification: for example, all Heralds after GA80000 (3 July 1962) will be to Mk2 chassis specification, but 1200 Mk1 chassis were still being built after early July and were given pre-GA80000 numbers (and were mostly Courier vans).

For convenience, changes are itemised in chronological order; the earlier cars in a range are invariably subjected to more alterations as faults are designed out, so this style is entirely logical. It does, however, have the disadvantage that the later cars, which have probably survived the years in greater numbers, appear to be short-changed in terms of coverage. To the many owners of 13/60s and MkII Vitesses I can but apologise, and ask them to read the references to earlier cars with care, as many of the changes which I describe will be equally applicable to their cars.

9

CHAPTER ONE

THE MODELS

ABOVE
A pair of early Heralds photographed in Eire during the press preview in March 1959. The coupé shows the first style roof fitted until June 1960. Pastel colours were a distinctive feature of many 948 Heralds, the duo-tone colour scheme being standard on coupés.

The decision to build the Triumph Herald onto a separate chassis was seen by many to be a retrograde step, as practically every small family car was of unitary construction by the mid-1950s, and Standard-Triumph had been using this method since the introduction of the Mayflower in 1949. The company, however, had been manoeuvred into a difficult position when its traditional supplier of bodyshells, Fisher and Ludlow, was taken over by the British Motor Corporation in 1953, and Leonard Lord of BMC saw no reason to offer continuing production facilities to a competitor. The only other independent company capable of a sizeable undertaking, Pressed Steel, was already committed to other customers, leaving Standard-Triumph with no option but to consider a design built up from several small suppliers' components. The separate chassis then provided a practical jig on which to assemble these components and also facilitated local assembly at Standard's plants throughout the world.

Standard-Triumph's designers, led by engineering director Harry Webster, incorporated several innovative ideas into the new model, despite the 'old-fashioned' basic concept. The chassis frame itself consisted of two main rails positioned as a central backbone, unlike the traditional British style of widespread main rails. To support the bodywork, light-weight out-

THE MODELS

LEFT
The 948 saloon was available to special order, equipped with the mechanical specification of the twin-carb coupé. The twin-carb saloon was never catalogued as a separate model, nor did it appear in the price list, but it was given a separate commission number series, and is now regarded as a distinct model. This early 1960 example was the author's first experience of Herald motoring, and is here seen in 1962. There are no external features to distinguish it from the ordinary saloon.

RIGHT
UK specification 948s featured plain painted bumpers, and white-wall tyres were not too popular on the home market. These late 1960 Heralds pose for the camera at a junction in the Midlands. The revised roof style fitted to the later coupés can be clearly seen. (BMIHT photo.)

ABOVE
The Herald was never a cheap car but Triumph tried hard to produce a 'bargain-basement' model in the Herald S. The bright side-trim was one of many features deleted from its specification, and the cheap one-piece radiator grille did nothing to disguise the austere look. Pale Yellow was not a popular colour choice.

riggers extended to the full width of the car. A sophisticated coil spring independent front suspension design was developed and linked to rack and pinion steering, while independent rear suspension by transverse leaf spring was the first attempt by a major British manufacturer at all-round independent suspension. Much has been made of the limits this placed on the car's roadholding but at the time this was considered a vast improvement on the live axle designs of its British competitors, and was seen to be more acceptable than similar suspension used on continental cars such as Volkswagen and Renault, which both suffered from the added burden of rear engine weight to upset the handling.

Giovanni Michelotti produced a bodywork design with crisp modern styling that was both elegant and distinctive. Using the freedom of the separate chassis, he produced a saloon which boasted a remarkable 93% visibility from the driver's seat, together with a huge bonnet, comprising the complete front bodywork, which hinged forward to give unrivalled accessibility.

Adjustable steering column designed to collapse on impact, 72 seating positions for the driver, crushable dashboard, folding rear seat giving extended luggage accommodation, fresh-air heating as standard – all were features which took the Herald out of the ordinary 'run-of-the-mill' family car market.

When launched in April 1959, the range consisted of just two models: a two-door saloon powered by the single-carburettor 948cc Standard Pennant engine, and a two-door coupé using a twin-carburettor version of the same engine. Pastel shades predominated in the range of colours offered, and a striking two-tone scheme with a pronounced wedge shape to the lower colour was available as an extra on the saloon, but standard on the coupé.

In December 1959, a twin-carburettor version of the saloon was introduced, incorporating all the mechanical features of

THE MODELS

ABOVE

Herald 1200 – the classic Herald saloon resplendent in gleaming Cherry and White with Matador Red interior. Behind is the coupé in Black and White, a relatively rare colour scheme. This photograph has been retouched to show the new Cherry colour introduced in 1963; earlier versions show the saloon in Signal Red with Grey upholstery, while the coupé is in Renoir Blue. The illustration, in fact, shows very early 1200s with HERALD lettering, the coupé retaining the large early sidelights, while the saloon is fitted with the new small sidelights.

ABOVE

The Herald 1200 Estate was a popular choice for market gardeners, and Triumph's Marketing Department cleverly identified this potential in its early publicity. Again, early cars are shown, the Signal Red car demonstrating the large sidelights, together with the front valance of all Mk1 Heralds. The Powder Blue car clearly shows the number plate shroud common to all estates.

13

COMPLETE GUIDE TO HERALD & VITESSE

ABOVE
This smart convertible is fitted with white wheels – a feature introduced on the 12/50, but standardised across the Herald range for a period during the early 1960s. Signal Red was by far the most popular convertible colour. The saloon visible in the background is in Gunmetal and Wedgwood Blue – one of the few duo-tones not to feature White as the second colour.

RIGHT
Conifer and Cactus was a popular choice for the desirable Herald 12/50. Note the sunroof, which was always supplied with Black fabric – colour co-ordination did not extend that far in the 1960s. This shot is particularly useful in showing the extent of the coverage of the second colour – cars were painted overall in the predominant colour, with the lighter colour overpainted on the exterior surfaces only.

the coupé, together with most of the trim features. Although never referred to in the sales literature of the time, this was a new model in its own right, carrying a unique commission number series which provides conclusive identification to the present-day owner.

The first completely new model was the convertible, introduced in March 1960. This incorporated all the distinctive features of the coupé, but used an entirely new rear body section with hood well and additional strengthening. Initially only available for export sales, it was to prove a popular variant in a market where convertibles were becoming rarer. Eventually the convertible was launched on the home market, but not until September 1960 – when the summer was nearly over!

Meanwhile, a major change was made to the coupé's roof in June 1960 when the original design was replaced with a three-piece construction with ribbed side-panels and an extended drainage gutter. Many thought the change was a visual improvement, but the main purpose was to make the roof more rigid and easier to produce.

In an attempt to break into the growing business and fleet market, a cut-price saloon, called the Herald S, was introduced in February 1961. The heater became an extra, and rubber mats replaced carpeting in the front footwells, although carpet was retained over the gearbox and in the rear compartment.

April 1961 saw the first major changes to the Herald, with the introduction of the 1200 range. The saloon, twin-carb saloon, coupé and convertible all ceased production, to be replaced by the same body styles but with a common mechanical and trim specification. White rubber bumpers were the obvious external identification, while inside a wooden dashboard, black switches and instrument faces (considered to be more stylish for the 1960s), and more thickly padded seats were introduced. The 1200s were generally much better cars than the original 948s. The new 1147cc engine produced much more torque, which meant that the axle ratio could be raised and still provide the car with better all-round performance – even though the S, which continued in production for two more years, could be ordered with the twin-carburettor engine, thus becoming both the cheapest and fastest Herald in the 1961-62 range!

One month after the announcement of the 1200 range, another new body style was introduced. The estate was to prove another popular variation on the basic theme, providing much more luxurious accommodation than other small estates, which tended to be downgraded into little more than a basic van with windows.

The Courier van was introduced in February 1962, using the basic style of the estate but with the trim levels of the S, including its one-piece radiator grille. This, however, was the least successful model, proving too expensive for the typical small retailer – the opposite marketing result to

ABOVE
The Courier was effectively an estate with metal rear side panels. Equipment and presentation was similar to the Herald S, with the latter's painted bumpers and plain radiator grille. Two chrome wipers were unusual on commercial vehicles of the time, but the twin wing mirrors fitted to Courier and estate models were necessary to meet legal requirements.

the estate. Sales struggled throughout its short life: a little over 5000 had been produced by its discontinuation in October 1964. Many remained unsold until converted into cut-price estates by fitting side windows and the optional rear seat – not an official conversion, but certainly undertaken before sale to original owners.

The next new model was to prove significant in more than one way, when a linered-down version of the six-cylinder Vanguard 6 engine was squeezed into the Herald. Leyland Motors, who had taken over Standard-Triumph in the spring of 1961, produced a model of quite distinctive character which created a market all its own. Installation of the new engine was not a simple exercise – virtually every mechanical component had to be reworked to cope with the additional weight and power. Not least of these components was the chassis frame itself, which emerged as a completely new unit, following the original design basis but different in nearly every detail. The main rails were altered from the original 3in square cross-section to 3¼in x2¾in to give greater rigidity, and increased space was provided around the gearbox to allow room for an overdrive. There was a much more substantial differential mounting, which also gave better support

THE MODELS

LEFT
Valencia Blue was a popular choice for the 13/60, the saloon shown here demonstrating the neat way Michelotti incorporated the single headlamp treatment into the Vitesse bonnet.

ABOVE
This lucky family is enjoying the benefits of one of the few family convertibles available in the late 1960s. White was one of the few colours which offered a choice of upholstery colour in the 13/60 range: Light Tan, Black or Matador Red could be specified, while the hood and hoodwell cover could be either black or white. Other than colour schemes, the potential purchaser of the 13/60 had no choices to make.

ABOVE & RIGHT

This 13/60 Estate looks especially smart in Royal Blue, displaying to good effect the high quality paint finish which Triumph achieved after the Leyland take-over. Even in 1967, Triumph was one of the few companies which offered a luxuriously equipped estate car, and benefitted from a loyal customer base in consequence.

THE MODELS

ABOVE
The new Vitesse 6 convertible – the V-shaped bonnet top was an easy identification feature from the front, but only the expert car-spotter would notice the alloy bumpers and wheel trim, when viewed from the side. The distinctive duo-tone colour scheme was unique to the model – all other Herald-based convertibles were supplied in monotone only.

for the rear body, and additional strengthening was incorporated at numerous other points. This new model, the Vitesse 6, was announced in May 1962. Available in both saloon and convertible body styles, it had high quality trim levels that were a major sales feature.

At the same time the revised chassis frame was also incorporated into the Herald 1200 range, although there was no external identification for the new Mk2 and it was never advertised as being a new model. The change has more significance today, as many body and chassis components are not interchangeable.

In March 1963, a further new Herald model appeared, with disc brakes and an uprated 1147cc engine producing 51bhp, in a standard saloon bodyshell topped by a smart folding sunroof. This 12/50 provided improved performance more cheaply than the Vitesse.

The last Herald S left the production lines in January 1964, followed in October by the last Couriers and coupés. The latter had been an extremely popular version in 948cc form, but by 1964 its appeal had faded as the Spitfire, introduced in 1962, took many of its potential customers.

Apart from relatively minor improvements to the Vitesse, the two ranges survived the next two years unchanged, while other Triumph models received their share of attention. Major changes were planned for 1967, however, when the Vitesse finally received the full 2-litre specification of the original six-cylinder design. External changes were limited to discreet new badges and a new rear number plate incorporating a reversing light. Drivers certainly noticed a difference, with over 30% more power and near 100mph performance. Introduced in October 1966, the new Vitesse 2-litre revived flagging sales in a dramatic manner.

Perhaps even more significant was the upgrading of the Herald in August 1967. By installing the 1296cc engine from the front-wheel drive Triumph 1300, a similar performance improvement was achieved, while the fitment of a single lamp version of the Vitesse bonnet and completely revised dashboard and trim provided a much greater visual impact. The 1200 in

COMPLETE GUIDE TO HERALD & VITESSE

the smooth six

TRIUMPH VITES[SE]

A member of STANDAR[D]

THE MODELS

RIGHT
The 2-litre Vitesse gave exhilarating performance for a small family saloon, and in MkII form provided the security of wishbone rear suspension. This Valencia Blue saloon represents the ultimate expression of the original Herald concept.

LEFT
Cactus Green was a new colour introduced in 1963, and is shown here with the optional duo-tone scheme incorporating a Black stripe, combined with Matador Red trim, on the Vitesse 6 saloon.

convertible and estate forms, together with the 12/50, were discontinued; only the 1200 saloon survived alongside the new 13/60, which was available in saloon, convertible and estate forms.

The revised frontal treatment of the 13/60, with three sets of horizontal bars across the radiator aperture, was extended to the Vitesse with the introduction of the new MkII in October 1968. However, the MkII's cosmetic changes – which included new badges, Rostyle wheel trims, anodised boot panel and 13/60 style dashboard – were only a minor part of the story. The big changes were to the rear suspension, which was revised to incorporate wishbone bottom links and variable-length drive shafts, using the rubber 'doughnuts' from the Triumph 1300. This eliminated the major wheel camber changes inherent in the original swing axle design, transforming the car's handling under extreme conditions. Engine modifications in the form of a revised camshaft and new cylinder head boosted power to 104bhp and top speed to a genuine 100mph.

After this last major revision, both the Herald and Vitesse were allowed to continue to the end of production with only minor specification changes, the last Herald 1200 leaving the factory in May 1970. Exactly 12 months later, the 13/60 and Vitesse were both quietly buried, by which time the Toledo was in full production and the exciting new Dolomite was about to appear. Components continued to be made for several more years, however, and it seems that Herald CKD kits – to the original Mk1/948 pattern, would you believe – were still leaving Coventry as late as 1973!

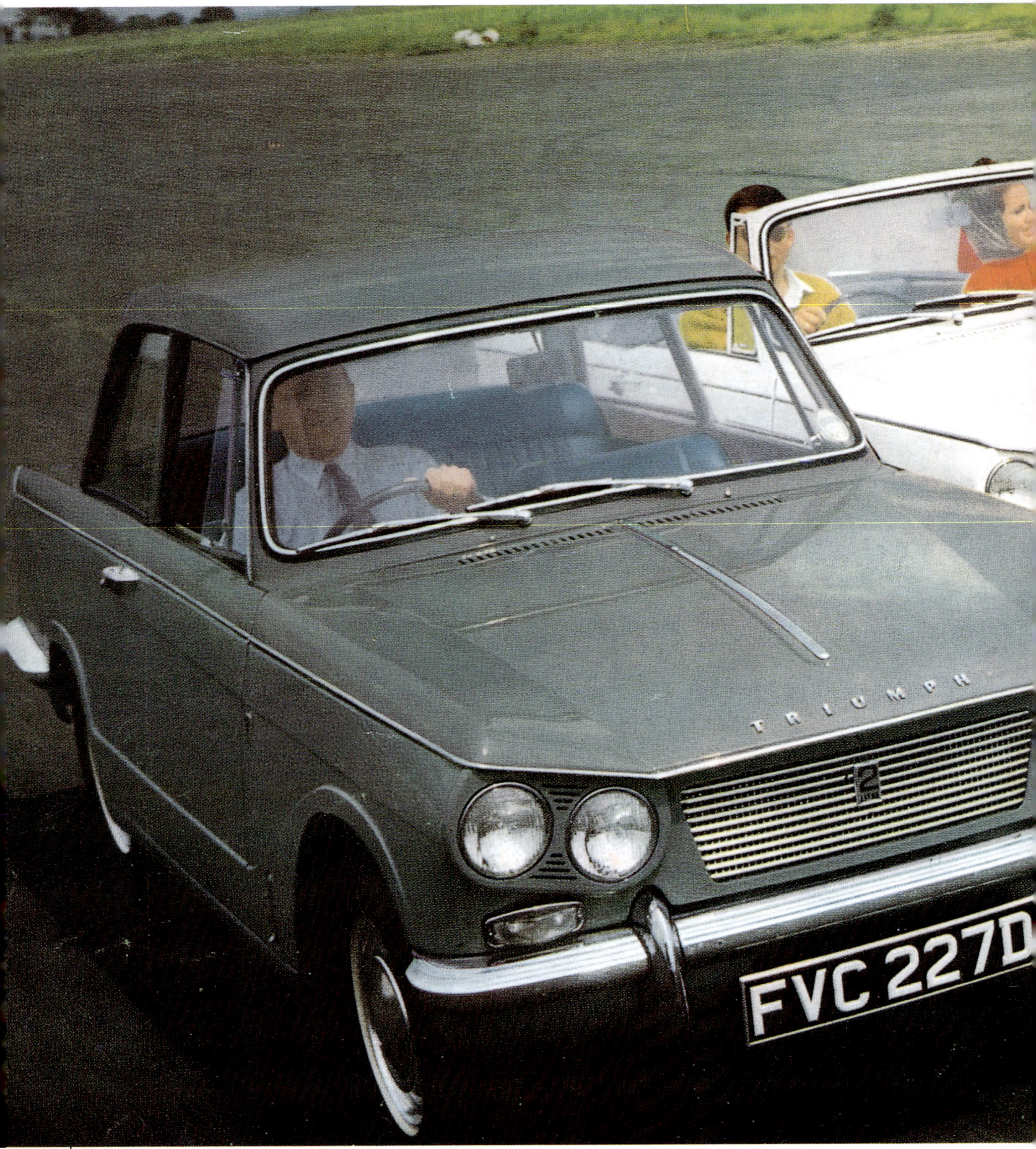
TRIUMPH
FVC 227D

THE MODELS

LEFT
External identification of the original Vitesse 2-litre is restricted to discreet '2-litre' badges on bonnet and boot; in all other respects, these late 1966 cars could be 1600s. The Signal Red saloon is equipped with the optional sunroof, a popular feature on many Vitesses.

COLOUR AVAILABILITY

White
Models **4, 5, 6, 7, 8, 9, 10**
Paint Code No 19
1961-71

Lichfield
Models **1, 2, 3, 4, 5, 6, 8**
Paint Code No 45
1959-64

Signal
Models **1, 2, 3, 4, 5, 7, 8, 9, 10**
Paint Code No 32
1959-71

Phantom
Models **1, 2, 3, 4, 5, 6, 8**
Paint Code No 38
1960-63

Royal Blue
Models **5, 6, 7, 8, 9, 10**
Paint Code No 56
1966-71

Sebring White
Models **1, 2, 3**
Paint Code No 29
1959-61

Coffee
Models **1, 2, 3**
Paint Code No n/a
1959-61

Cherry
Models **5, 8, 9**
Paint Code No 22
1963-68

Gunmetal
Models **5, 7, 8, 9, 10**
Paint Code No 18
1963-69

Monaco Blue
Models **1, 2**
Paint Code No n/a
1959-60

Black
Models **1, 2, 3, 5, 8**
Paint Code No 11
1959-68

Sienna
Model **10**
Paint Code No 23
1969-71

Pale Yellow
Models **2, 3, 4, 5, 8**
Paint Code No n/a
1960-62

Dolphin
Models **5, 8**
Paint Code No 48
1965-68

Wedgwood
Models **5, 7, 8, 9, 10**
Paint Code No 26
1963-71

Cactus
Models **5, 8**
Paint Code No 15
1963-66

Alpine Mauve
Models **1, 2, 3**
Paint Code No n/a
1959-61

Jonquil
Models **5, 8**
Paint Code No 14
1963-64

Slate
Models **7, 9, 10**
Paint Code No 68
1968-71

Powder Blue
Models **1, 2, 3, 4, 5, 8**
Paint Code No n/a
1959-63

Conifer
Models **5, 7, 8, 9**
Paint Code No 25
1963-68

Targo Purple
Models **1, 2**
Paint Code No n/a
1959-60

Jasmine
Models **5, 7, 9, 10**
Paint Code No 34
1967-70

Valencia
Models **5, 7, 9, 10**
Paint Code No 66
1967-71

Olive
Models **5, 7, 8**
Paint Code No 35
1963-66

Damson
Models **7, 9, 10**
Paint Code No 17
1968-71

Saffron
Models **5, 7, 10**
Paint Code No 54
1970-71

Renoir
Models **5, 6, 8**
Paint Code No n/a
1961-64

Notes to table

Model codes

1 - 948 saloon
2 - 948 coupé
3 - 948 convertible
4 - 948S
5 - 1200 & 12/50
6 - Courier
7 - 13/60
8 - Vitesse 1600
9 - Vitesse 2-litre
10 - Vitesse Mk II

CHAPTER TWO

Colour Schemes

From mid-1960s, Commission Number Plates include paint/trim code numbers, which provide positive evidence of a car's original colour scheme.

Herald 948s and 1200s were available in duo-tone (saloon and coupé only), normally of body colour plus white centre section. Exceptions were: 948 coupés – Black duo-toned with either White or Alpine Mauve; 1200s – Conifer and Olive duo-toned with Cactus, Black with White or Cactus, Gunmetal with Wedgwood.

Vitesse 1600s were duo-toned on both saloon and convertible with a contrasting side-stripe. This was normally White, except for White itself, which had a Black stripe; later Black cars (from June '63) had a Cactus stripe, as did Olive and Cherry; Wedgwood and Cactus had a Black stripe.

With the exception of 948 coupés, duo-tones were always a special order option, mono-tone coupés being also available as a no-cost option. The popularity of the duo-tone colours declined during the 1960s, and by 1966 were such a rare choice that the option was deleted. The 13/60 and Vitesse 2-litre models were not available with a factory duo-tone. The option was also never offered on 948 convertibles, nor on 1200 convertibles, estates, Ss and Couriers.

COLOUR COMBINATIONS

Model	Upholstery	Paint
Herald 948	Alpine Mauve	Alpine Mauve, Monaco Blue, Black/Alpine Mauve
	Coffee	Black, Coffee
	Phantom Grey	Black, Lichfield, Signal Red, Alpine Mauve, Phantom Grey, Powder Blue, Monaco Blue
	Matador Red	Black/White, White, Lichfield, Phantom Grey
	Targo Purple	Targo Purple
	Black	White, Red, Powder Blue
Herald 1200	Matador Red	Black, White, Cactus, Conifer, Lichfield, Cherry/White, Gunmetal, Phantom Grey
	Phantom Grey	Lichfield, Phantom Grey, Powder Blue, Renoir
	Black	Black/White, White, Lichfield, Signal Red, Pale Yellow, Jonquil, Saffron, Powder Blue
	Coffee	Black
	Midnight Blue	Wedgwood, Gunmetal
	Cactus Green	Black/Cactus, Conifer/Cactus, Olive/Cactus, Cherry
	Light Tan	Royal Blue, Valencia Blue
	Shadow Blue	Gunmetal, Dolphin, Royal Blue, Valencia

Note to table
Paint/trim combinations are numerous, with considerable variations through the years, as well as between models, but these details are believed to be comprehensive – although it should not be assumed that all combinations were available throughout the model's existence.

25

Model	Upholstery	Paint
Herald 13/60	Black	White, Signal Red, Jasmine, Saffron, Royal Blue, Conifer
	Matador Red	White
	Light Tan	Damson, Valencia, White, Olive
	Shadow Blue	Gunmetal, Slate, Wedgwood, Royal Blue, Valencia
Vitesse 1600	Phantom Grey	Lichfield, Powder Blue, Phantom Grey, Renoir
	Matador Red	White, Black, Cactus, Conifer, Lichfield, Phantom Grey, Dolphin
	Black	Jonquil, Signal Red, White, Black, Powder Blue, Pale Yellow
	Midnight Blue	Wedgwood, Gunmetal
	Cactus	Olive, Conifer, Cherry
Vitesse 2-litre	Black	White, Signal Red, Jasmine, Wedgwood, Royal Blue
	Matador Red	Black, White, Conifer
	Light Tan	Valencia
	Cactus	Cherry
	Midnight Blue	Gunmetal, Slate Grey, Wedgwood, Royal Blue
	Shadow Blue	Valencia, Gunmetal
Vitesse MkII	Black	White, Signal Red, Damson, Sienna, Slate, Jasmine, Saffron, Wedgwood, Royal Blue, Valencia
	Matador Red	White
	Light Tan	White, Signal Red, Damson, Sienna, Valencia
	Shadow Blue	Gunmetal, Slate, Royal Blue

CARPET COLOURS

Model	Carpet
Herald 948	Alpine Mauve, Coffee, Phantom Grey, Matador Red or Targo Purple to match upholstery colours up to February 1960 (G31008/Y6403). Then Black or Charcoal to December 1960 (G59934). Then Charcoal Grey only.
Herald 1200	Charcoal Grey only to December 1967 (GA236219). Then Black only.
Herald 13/60	Black, Matador Red, Light Tan or Shadow Blue to match upholstery colour.
Vitesse 1600	Smoke Grey/Black mottle only.
Vitesse 2-litre	Black only.
Vitesse MkII	Black, Matador Red, Light Tan or Shadow Blue to match upholstery colour.

CHAPTER THREE

BODYWORK

ABOVE
The original bonnet style used on all 948s and 1200s featured horizontal wing tops flanking a gently sloping top panel. This early 1200 features a type of sidelamp used only for an 18-month period; side and flasher functions were provided by one double-filament bulb.

ABOVE
The Vitesse introduced the 'Chinese-look', with this pronounced V-shape to the leading edge of the bonnet. Four headlamps were all the rage in the early 1960s, but Michelotti's treatment was one of the most successful conversions of an existing style.

There was remarkably little change in general specification to Michelotti's original body design, and most of the alterations made during the 13 years of production relate to the fixtures and fittings (dealt with in later chapters).

The most numerous changes were made to the bonnet, which was made in three basic styles: 948 and 1200 Herald, 13/60 Herald, and Vitesse. The 948/1200 bonnet's only external change was to the grille on the top panel, in front of the windscreen. Up to 1963, the bonnet top panel had a large rectangular hole with a separate grille spot-welded to the underside, but from late 1963 this design was changed to incorporate a series of slots punched into a depression on the main top panel. This style also features on all Vitesses, although now the slots were in two separate depresssions with a central rib continuing between them. The 13/60 used the same top panel as the Vitesse, so had the same feature.

The story is not so simple on opening the bonnet. Early 948s featured a complex inner wing pressing, to which a rubber shield was rivetted to protect the engine bay from road spray. In November 1959,

RIGHT
The Vitesse bonnet, converted back to single headlamps, provided a simple revision to the basic Herald 1200 style for the new 13/60.

BELOW
On early Heralds the grille on the bonnet top was of this pattern – a separate panel spot-welded to the underside of the bonnet.

ABOVE
Herald 1200s from 1964 were fitted with a bonnet incorporating this revised grille, now integral with the bonnet panel.

RIGHT
The 'slant-eyed' bonnet fitted to Vitesse and 13/60s had two grille panels either side of an extended centre rib.

RIGHT
Early 948s were fitted with this pattern of bonnet location, the bulge on the bonnet pressing resting on the fabric-covered 'flat' V-shaped bulkhead bracket.

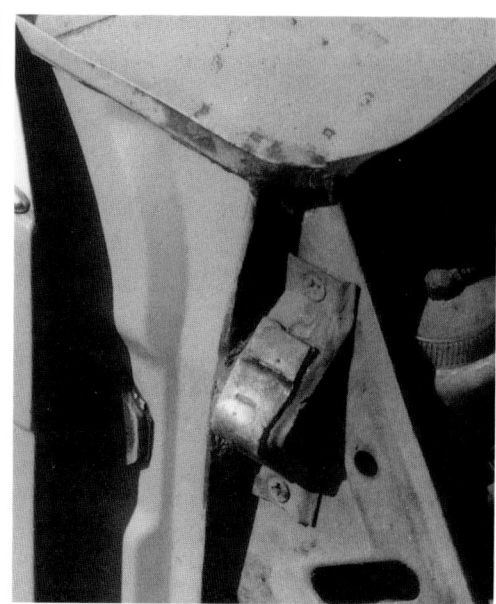

LEFT
In 1960, the bonnet pressing gained this extra swelling, providing more positive location onto the new concave V-shaped bracket.

separate metal panels, mounted onto the chassis frame and suspension turrets, were introduced, so that the inner wing was simplified in shape. There was a short period between the fitment of the metal panels and the deletion of the rubber shields. After November 1959 (G15548), however, the simple inner wing style was standardised, and continued unchanged to the end of production in 1971.

Locating the bonnet onto the bulkhead must have caused Triumph some problems, as there were no fewer than four designs in the first few years of production. The first style featured a wedge-shaped pressing retained to the bulkhead by two setscrews, with a length of webbing stapled on its surface, and mated to a matching wedge welded to the bonnet. In October 1960 (G58649), this was replaced by a similar arrangement, but with a groove in the bulkhead pressing and a corresponding bulge on the bonnet pressing. This arrangement continued through to mid-1962, and was featured on early Vitesses. Around June 1962, an adjustable rubber cone and corresponding socket was introduced. This design was first used in the manner familiar to Spitfire owners, having the cone mounted onto a bracket on the bonnet and the socket on the bulkhead. Unlike the Spitfire, however, this was reversed in August 1962, so that the cone was located onto the bulkhead, with

LEFT
First attempts with the adjustable rubber cone had the socket on the bulkhead.

RIGHT
From 1963 the bonnet location was standardised with the rubber cone now located on the bulkhead.

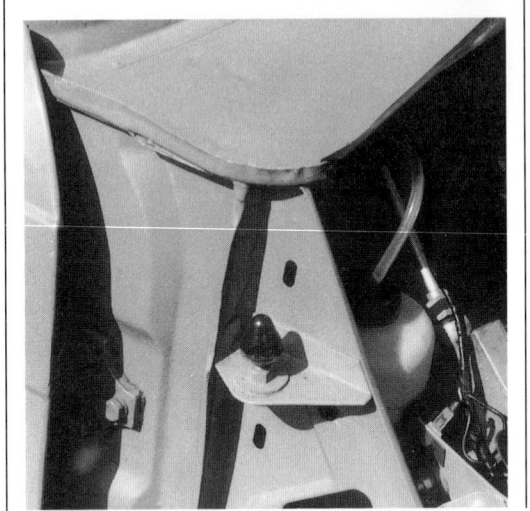

RIGHT
Early Heralds had a spring to counter balance the bonnet, fitted between the tubular hoop of the bonnet, and a bracket on the top edge of the radiator. The rod from the upper wishbone mounting to the bonnet pivot relied on slotted holes for adjustment.

RIGHT
Later cars had this arrangement of tie rod, bonnet stay and counterbalance spring. The adjustable tie bar was introduced early on in the life of the 1200, the bonnet stay being incorporated from October 1961 (GA35950). This arrangement was common to all Vitesses.

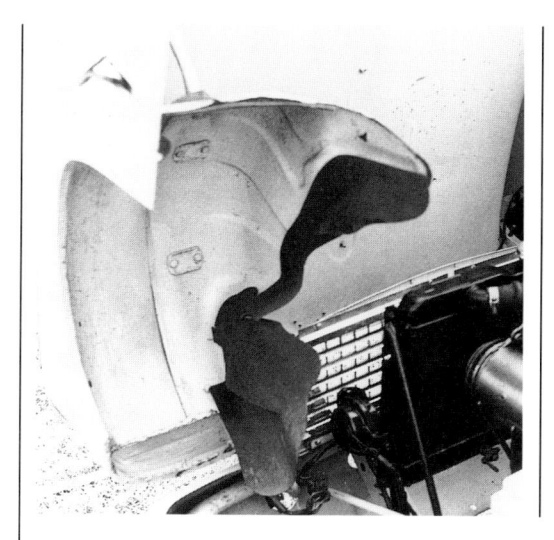

BODYWORK

LEFT
The inner wing of early 948s had this complex pressing, with the rubber curtain riveted in position.

RIGHT
All models from late 1959 were fitted with this simplified inner wheel arch.

RIGHT
The top corner of the door reveals the only change made to this panel. The early unit matches the bulkhead curve better, but it was probably more difficult to fit the outer skin at this point.

the socket now on the bonnet – this revision remained for the balance of production.

The doors appear unchanged from the earliest Herald to the last Vitesse, but in fact in July 1959 (G3676) one small change was made, a detail so insignificant that one wonders why it was made – especially since it necessitated new pressings for both inner door and outer skin. On early cars the top edge of the door continues in a curve around the bulkhead at the extreme front corner, but on later cars there is a straighter edge, which no longer matches the line of the bulkhead.

The two main body sections, the bulkhead and the rear tub, differed between models, and also featured alterations over the years. The only difference to the bulkhead between models is to the top panel above the windscreen: the convertible models were unique in having tapped blocks inserted into the double skin in order to fit the hood fixing sockets. All fixed head

31

RIGHT
The saloon roof style remained unchanged from the first Herald to the last Vitesse. The only detail change was the deletion of the rain guttering across the windscreen in 1961 (see also the following chapter). All cars from November 1959 used a common boot pressing with a completely smooth edge to the recessed panel.

RIGHT
The saloon roof reaches to within 2in of the boot lid, whereas the shorter coupé roof leaves 12in of rear deck.

RIGHT
The convertible rear deck is significantly shorter than the coupé's, which unfortunately means that a coupé roof cannot be used as a convertible hard-top.

BODYWORK

RIGHT
Early coupés were fitted with the original Michelotti roof design with the plain rear quarters. Note how the rain guttering follows the profile of the window.

BELOW
The estate roof followed the saloon style, with fixed single side windows. Rain guttering was added above the rear door.

ABOVE
The second style of coupé roof featured these ribbed quarter panels and revised guttering. The same style was fitted to 1200s, but with the same revision at the front as the saloon.

models had a common top flange here. The rear tub came in various versions: three different lengths of top deck for saloons, coupés and convertibles (the latter also incorporating the hood well and additional stiffening), while the estate and Courier, of course, had no top deck at all. However, there were two significant changes related to age.

Completely new floor panels were needed to both main body sections as a result of the new Mk2 'Star' chassis, introduced with the Vitesse and incorporated into the Herald range at GA80000 (July 1962). Consequently, there are Mk1 and Mk2 versions of all the Herald body styles, and anyone hoping to locate a better unit during a rebuild should bear this in mind. Mk2 bodywork can be fitted to a Mk1 car rather more easily than Mk1 bodywork to a Mk2 car – but neither exercise is straightforward.

The second bodywork change affects late 13/60s and Vitesses only: in 1970 the wiper mechanism was revised to use fwd Triumph 1300/1500 components, and this necessitated an altered bulkhead with the wiper arms being more widely spread. The easiest way to identify the later version is by the location of the left-hand wiper, which was now positioned outboard of the left-hand washer.

The roof panel obviously differed between the various body styles, but in addition there were three distinct coupé roofs and two saloon roofs. The original coupé was designed with plain rear quarters, and a rain gutter which followed the line of the side windows. In June 1960 (Y12433), a revised design was introduced

33

RIGHT
The Courier used the same roof pressing as the estate, with these metal side panels welded to form a complete one-piece unit. A series of bolts located the roof to the upper wing structure.

BELOW
Early 948s were fitted with a boot lid incorporating a slight curve to the top of the recessed panel.

ABOVE
The leading edge of the roof on all but the 948 Herald – no bright trim on the guttering and a horizontal flange were common to both Herald and Vitesse. This shot also shows the bright windscreen trim introduced with the 1200.

which featured guttering sweeping back to the bottom of the roof, while the triangular side panels featured horizontal ribbing. The rain gutter on both these versions, and also on the original saloon roof, continued across the top of the windscreen. With the introduction of the 1200 model, this front guttering was deleted and the roof now finished in a horizontal flange. There were no further alterations to the roof.

The different lengths of roof for the various body styles required changes to the top deck and upper wing panels of the rear body unit, but the general design remained unchanged. The boot, however, did receive modifications over the years: the earliest panel featured a depression under the upper lip of the rear panel to take a recessed number plate illumination lamp. From November 1959 (G17506), this was replaced by a lamp mounted onto a separate cowl, which required a boot lid with smooth contours. The Vitesse 1600 and 13/60 featured a number plate lamp which doubled as a boot lamp, and so a rectangular hole was punched in the underside of the lip, although the overall shape of the panel was unchanged. A common boot stay was used on all models, but the bracket was strengthened in May 1965 (GA178000).

The front valance fitted to all Mk1 Herald chassis was of a common design, having a smooth underside across the full width. With the introduction of rubber bumpers on the 1200, retaining flanges were spot-welded in position, and the outer corners of the 'bumper' were given a flatter profile. The Vitesse, with its forward-mounted radiator, had a central bulge and ventilation slot on the valance underside, which was incorporated onto the Herald with the introduction of the Mk2 'Star' chassis, although the upper surface necessarily

BODYWORK

RIGHT
An integral grille was added to the heater intake in 1962. Earlier cars, fitted with the large entry holes, suffer from a build-up of dust which blocks off the heater.

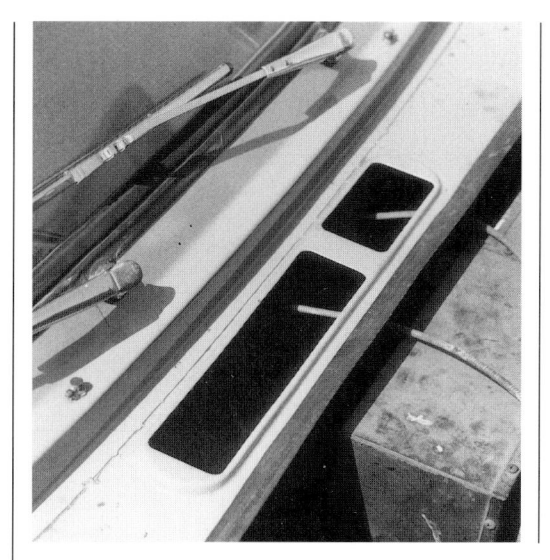

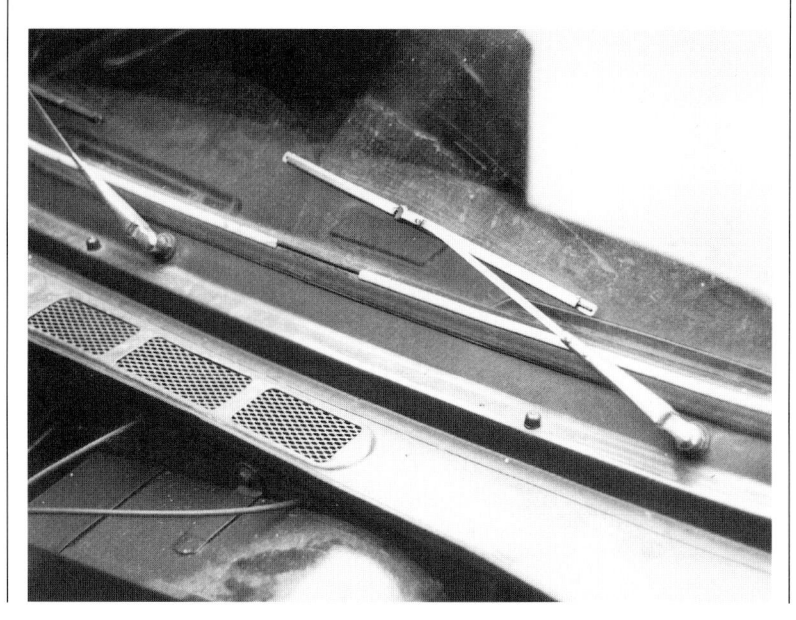

LEFT
These two shots show the change to the front bulkhead – the widely spaced wiper arms were introduced to facilitate the fitment of the 1300-sourced components.

RIGHT
The Weathershield roof was a standard fitting on the 12/50, and a popular option on the Vitesse and 13/60. The tinted perspex Galemaster deflector was a practical extra which reduced buffeting at speed.

RIGHT
Those Heralds not fitted with rubber bumpers featured a front valance with this bulbous outer corner. Very early Vitesse 1600s also featured this shape, although hidden behind the aluminium cappings.

RIGHT
Rubber-bumpered cars were fitted with a front valance which had a much flatter outer corner (the bumper retaining flange is missing on this car).

ABOVE
The two designs of handle – the later design, in the lower picture, provided more leverage and facilitated operation.

RIGHT
When the flatter shape was fitted to the Vitesse, this bracket was spot-welded in position to provide an anchor for the outer corners of the cappings.

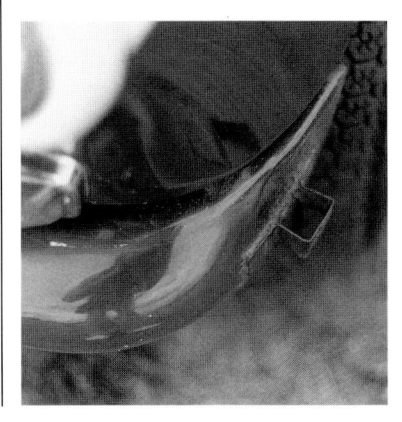

BODYWORK

LEFT
Heralds fitted with the Mk1 chassis featured a smooth underside to the front valance – the number-plate was fitted to brackets bolted through the valance.

RIGHT
Introduced on the Vitesse to provide room for the forward-mounted radiator, a front valance incorporating this air intake was fitted to all Mk2 chassis. This Courier is fitted with the panel supplied as a replacement in later years – originally it would have had the early-style corner.

retained the flanges on which the radiator grille was mounted. The earliest Vitesses featured the same bulbous corners as the earlier Heralds, but this was very soon modified to the flatter shape, with the addition of a bracket on which the aluminium bumper trim could be mounted. The 13/60 valance was similar to that of the Vitesse, but had a stepped-up central top surface, requiring a similarly stepped mounting bracket.

The rear valances on 948 Heralds were all welded integrally with the body, but replacement panels were always supplied with bolted flanges to ease repair. With the introduction of the 1200, the opportunity was taken to revise the initial construction of the bodywork to incorporate these bolted valances – although the S kept the welded panels.

All Heralds from the introduction of the 1200, except the Herald S and Courier, were fitted with moulded white rubber bumpers as standard, the outer ends being closed off by alloy pressings. All Vitesses used a standard pattern of extruded aluminium bumper, highly polished and anodised, and rivetted in position.

The sunroof fitted as standard to the Herald 12/50 from its introduction in March

37

RIGHT
All vehicles were fitted with the same pattern of rear valances, although 1200s and 13/60s had spot-welded flanges to retain the rubber bumpers.

RIGHT
The convertible hood had a large rear window and additional corner windows, providing much better rearward visibility than usually found on convertibles.

1963 was of a folding pattern of metal framework covered by black leathercloth externally, and lined internally with the same white material as was used for the standard roof lining. It was opened by a symmetrical white handle located centrally at its leading edge. Very late 12/50s were fitted with a revised handle, which was considerably larger than the original, and was now asymmetrical to give better leverage, making the operation of opening easier. This change appears to have occurred in 1967, and so is more common on 13/60s and Vitesses, where these were ordered with the optional factory-fitted sunroof.

The hood fitted to convertibles remained unchanged throughout the model's life: all frames, brackets and fittings are common from the earliest 948 to the last Vitesse. The 948 convertibles, however, had the hood frame painted in Wisteria (a pale pink colour). Herald 1200, Vitesse 1600 and 2-litre frames were painted in Phantom Grey. Herald 13/60 and Vitesse MkII frames were painted Black. The hood covering was available in black or white to choice, but 948s were unique in having a white inner material surface with the outer colour black. The header rail was covered with material to match the framework – white on 948s, grey on the majority of cars, and black on 13/60s and Vitesse MkIIs.

CHAPTER FOUR

EXTERIOR TRIM

LEFT
Initially this stainless strip was used only on 948 coupés, but subsequently appeared on all 948 convertibles and early 1200s.

A large chrome bonnet handle was a distinctive feature of early Heralds, and was fitted to all 948s other than the S, and on early 1200s up to June 1961 (GA 15938). The 948 saloons, in both single and twin carb form, used a small chrome foot at the leading edge of the handle, while the coupé, followed by the convertible and all early 1200s, featured a stainless strip running from the handle towards the front edge of the bonnet. When the handle was discontinued, this strip was extended back to the air vent at the rear of the bonnet; on all Vitesses and 13/60s the strip was extended further to divide the vent in two.

Door handle design remained unchanged throughout production, although a change in the door lock mechanism occurred in February 1962 (GA 50248) when the door button was incorporated into the handle mechanism. Before this, the button was part of the lock, the handle being completely separate. All 948s, and 1200s before this date, featured key-locking on both driver and passenger doors, but all subsequent models had a plain button on the passenger door, which could only be locked from the inside of the car. There were no alterations to either the boot handle or the bonnet catch assembly, although the boot handle was sourced from both Wilmot Breeden

RIGHT
The chrome bonnet handle used a small foot at the leading edge on 948 saloons. The thick rear edge disguised the two studs used to clamp the bonnet's tubular framework.

RIGHT
After June 1961, the bonnet handle was discontinued. The tubular framework was now retained by two bolts, and the stainless strip extended to cover them.

39

RIGHT
When fitted to the Vitesse, and subsequently to the 13/60, the stainless strip continued to the rear of the bonnet, dividing the air vent into two separate grilles.

ABOVE
Early door locks incorporated the button as part of the internal mechanism, with a separate handle surrounding the button. From February 1962 the button was incorporated into the handle, with an adjustable setscrew to engage with the internal lock. All non-locking passenger handles were of this construction.

LEFT
This boot handle was common to all cars, from 948 Herald to MkII Vitesse.

EXTERIOR TRIM

RIGHT
The chrome-plated handle on each side of the bonnet carried an 'M' for Michelotti.

ABOVE
Estate cars and Couriers were fitted with a pair of handles on the rear door – a non-locking handle on the left, locking on the right. They are here shown in the open position; each turns towards the outer edge of the car to close.

RIGHT
The two styles of boot hinge: above, the style fitted to the majority of 948s; below, the later version with a raised portion directly above the hinge-pin.

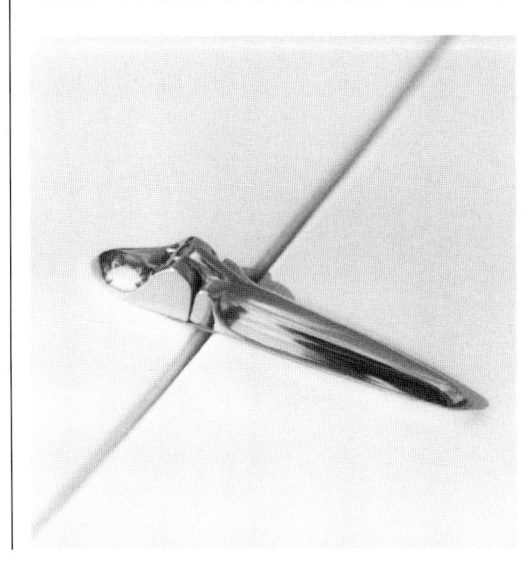

and Steiber, the lock barrels of these alternatives not being interchangeable. The boot hinge, however, showed a minor change from March 1961, when a small pip was added to the hinge in order to restrict the opening. Before the introduction of the 1200, the hinge had provided no stop to limit the opening of the boot, so if the check strap should break or be removed, the boot could be opened wide enough to touch the rear window.

All models except Herald S and Courier featured stainless strips down the length of the bodywork. Those on Heralds finished in front of the petrol cap on the nearside and in a corresponding position on the offside, while the Vitesse had an additional piece of trim behind the petrol cap and a correspondingly longer 'spear' on the driver's side. The 948 Heralds featured additional brightwork around the roof guttering. Cars before May 1960 (G36016/Y7956) used a single piece across the top of the screen with separate side pieces, the junction being on the front corners of the screen pillars. After this date, a four-piece trim set was used, with joints at the centre of the windscreen and just in front of the rear edges of the door quarter lights. In addition, 948 coupés and convertibles featured a similar piece of trim mounted on the bonnet, surrounding the radiator grille.

41

RIGHT
Heralds (except S and Courier models) were fitted with stainless side trim which stopped short some distance forward of the rear lights.

RIGHT & BELOW
All Vitesses featured these full-length 'spears'. On the nearside an additional trim was added behind the petrol filler.

BELOW
At the front the side trim was continued across the leading edge of the bonnet on all Vitesses and 13/60s, linked at the corners by small castings retained by a single screw.

EXTERIOR TRIM

BELOW
The 948 Heralds featured bright trim around the roof guttering. Early cars had a one-piece front strip with separate side-pieces, the joint being on the forward corner as here.

LEFT & BELOW
On the left, the bonnet grille surround on 948 coupés and convertibles; all other cars with the traditional Herald bonnet were left unadorned (below).

ABOVE
From May 1960 the roof trim was revised, with the front strip now divided at the centre, and the outer joints moving round to a new position above the quarterlights (see page 54).

The original 948 headlamp cowl continued through the production of the 1200, although a minor revision occurred in May 1963, when the depth of the peak was reduced from 3¾in to 3¼in. With the change of style used on the Vitesse, and subsequently on the 13/60, a complete revision of headlamp surrounds was required. The stainless side moulding was continued across the front of the bonnet, linked at the corners by chrome-plated castings. The actual headlamp surrounds on the Vitesse were painted and originally featured slots between the individual lamps; these slots were deleted soon after the introduction of the MkII, later cars having plain surrounds. The 13/60s used a chrome-plated surround with matt black painted infills at the top outer corners.

At the rear, 948 Heralds used a one-piece chrome-plated brass pressing to surround the rear light clusters; all subsequent models used a common design featuring a V-shaped chrome-plated casting above a pair of stainless strips.

The original 948 radiator grille design, using two alloy pressings mounted behind a painted steel panel, remained unchanged when fitted to the standard 1200 models. The S used a cheaper one-piece alloy pressing with large square apertures, mounted on the front face of the normal

43

LEFT & RIGHT
The original Herald headlamp cowl measured 3¾in front to back (left); from mid-1963 this peak was reduced to a depth of 3¼in (right).

BELOW
When viewed from underneath, the difference is more apparent. The early version on the right has a more pronounced cut-away on the underside, which can be identified by touch when in situ.

BELOW
All Vitesse 1600s and 2-litres, together with early MkIIs, were fitted with headlamp surrounds incorporating these slots; when a duo-tone colour scheme was specified, these were painted to match the side stripe. This shot also shows the ribbed aluminium bumpers fitted to all Vitesse models, and retained by rivets through the valance panels.

BELOW
Most Vitesse MkIIs were fitted with plain pressings without the slots used on earlier cars.

EXTERIOR TRIM

LEFT
Herald 13/60s were fitted with this chrome-plated surround, incorporating a small triangular panel painted matt black.

RIGHT
In the upper picture, the one-piece chromed brass trim of the 948. Note the sharp top point and thinner upper section compared with the three-piece construction of later cars (below).

steel panel; the Courier used the same alloy pressing, mounted on a light steel frame and supported at the top by brackets back to the radiator, rather than bolts forward to the overrider as on the standard grille. The 12/50s used a grille mounted in the same manner as the Courier grille, but featured a more complex pressing with a finer mesh, and incorporating a central dividing strip. Vitesse 1600 and 2-litre models used a one-piece pressing, similar in style to the 12/50, mounted behind the front panel of the bonnet itself. The 13/60s introduced a new style of grille consisting of three sets of bars, produced as a one-piece alloy casting and mounted onto the outside of the bonnet front panel. This style was continued for the MkII Vitesse, although the casting was narrower to fit between the twin headlamps. In June 1970 (GE69234), the 13/60 grille was changed to a plastic moulding, but the Vitesse retained its alloy unit to the end of production.

Only two styles of front overrider were used: the tall Herald version on all models fitted with the traditional bonnet style, and the smaller style used initially on the Vitesse, but also fitted to the 13/60. The rear overrider was a common fitting for all models. In all cases the overrider is a chrome-plated steel pressing incorporating welded mounting flanges and captive nuts.

All 948cc models featured painted 'bumpers', which were in fact part of the

45

RIGHT
Herald 948 and 1200 radiator grille consisted of two pressings, located to the back of the painted frame by a series of set screws. Note the Triumph badge fitted in the radiator grille area on pre-13/60 Heralds.

ABOVE
Herald S and Courier (upper pictures), and 12/50 (lower picture), had a one-piece grille pressing.

valance pressings. On all 1200s and 13/60s, except the Courier, these were covered by white rubber strips clipped over spot-welded flanges and retained at the outer ends by alloy pressings screwed into the valances. All Vitesses used a common style of aluminium trim with a series of horizontal ribs on the upper surface; these were riveted through the valances. This was the first time that such extensive use of extruded aluminium for bright trim had been made in the motor industry.

A chromed and painted badge, originally of cast alloy and subsequently of plastic, was fitted to the radiator grille of 948, 1200 and 12/50 Heralds, the same badge appearing on the top bonnet panel of the Herald S and Vitesse 1600. The 948 cars had the letters 'TRIUMPH' across the bonnet top, as did all Couriers, Vitesse 1600s, Herald 1200s after October 1962, and 13/60s. The Herald S, and 1200s before October 1962 (GA90334), had the letters

EXTERIOR TRIM

LEFT
Vitesse 6 and 2-litre models used a grille pressing with the fine mesh of the 12/50, mounted onto the rear of the bonnet front panel. Note also the smaller overrider always used with the Vitesse-style bonnet, and the style of TRIUMPH lettering and bonnet badge unique to the Vitesse 6 model.

RIGHT
This revised grille was introduced on the 13/60; black fabric was stuck between the individual bars.

ABOVE
The Vitesse MkII grille was identical to the 13/60 in all respects except overall width.

RIGHT
The tall front overrider fitted to those vehicles using the traditional bonnet style; cars with the Vitesse-style bonnet had a full-width front panel, so a smaller overrider was fitted.

47

LEFT
The rear overrider fitted to every Herald and Vitesse.

RIGHT
This alloy pressing was fitted to the outer extremities of the Herald rubber bumpers.

ABOVE
All Triumph identification was deleted from the fron of the Vitesse MkII. This badge was mounted on the bonnet top (left), while the side badges were upgraded to read 'Mk2' (right).

LEFT
The Herald S combined HERALD lettering with the Triumph badge. The Vitesse 6 shared this layout but used TRIUMPH lettering with the badge.

48

EXTERIOR TRIM

ABOVE
The Vitesse 2-litre dispensed with the Triumph badge, using this 2-litre badge on the radiator grille and bonnet side.

ABOVE
All Triumph identification was deleted from the front of the Vitesse MkII. This badge was mounted on the bonnet top (left), while the side badges were upgraded to read MkII (right).

ABOVE
The Leyland logo fitted to 1971 13/60s was positioned adjacent to the bonnet catches.

'HERALD' in this position. The Vitesse 2-litre had no identification on the bonnet top, but a 2-litre badge appeared in the centre of the radiator grille. Vitesse MkIIs had a new style badge in cream and chrome mounted off-centre on the bonnet top. Vitesse 2-litre models had additional badging on the bonnet sides, reading '2-litre' or 'MkII' as appropriate. The very last 13/60s had self-adhesive 'Leyland' logos fitted low down, adjacent to the bonnet catches.

Early 948 Heralds – those fitted with the recessed number plate light – had TR3A-style 'TRIUMPH' lettering mounted below this light. All other cars except 13/60s and Vitesse MkIIs were fitted with the bonnet 'TRIUMPH' lettering on the upper boot lid surface. Couriers had the full model title spread across the tailgate. The 13/60s and Vitesse MkIIs used the Triumph 1300-style 'Triumph' badge to the right of the number plate – supplemented by 'Herald 13/60' or 'overdrive' where appropriate. Earlier

49

RIGHT
The Triumph lettering fitted to early Herald 948 bootlids; this was only fitted to 1959 saloons and coupés.

RIGHT
The 1959 coupés carried additional badging at the rear: Coupé script on the bootlid and Herald script on each rear wing.

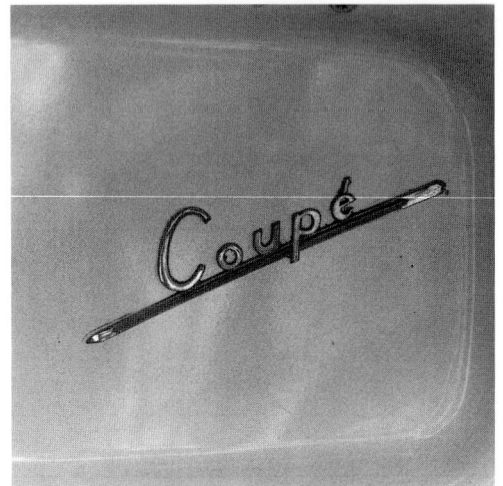

ABOVE & RIGHT
Later Herald 948s carried the TRIUMPH lettering on both bonnet and boot, but no other badging was used except the crossed flags motif fitted to the rear wings of coupés and convertibles. These motifs were also fitted to 1200 coupés and convertibles, and most 13/60 convertibles.

EXTERIOR TRIM

LEFT
These two badges identified all Herald 1200 models (left), the 1200 being replaced by S on Triumph's bargain basement model (right).

RIGHT
Early 12/50s carried their special badge on the boot and rear wings, but the last cars lost their identity from the side.

ABOVE
The Courier carried its name in letters across the tailgate, but no further badging in the number-plate recess.

Vitesses had the model name in script in this position, supplemented by either a '6' or '2-litre' badge. When the optional overdrive was specified, this was proclaimed by an 'overdrive' script mounted on the left-hand side of the boot on Vitesse 1600s, and by a more discreet badge under the 'Vitesse' script on 2-litre cars. On some non-overdrive cars, both 1600 and 2-litre, an underline was fitted beneath the script – but this was not always applied to either model.

Herald 1200s were fitted with a pair of badges identifying the model, while the S used the same Herald badge, but its own 'S' logo in place of the '1200' badge. The 12/50s also used the Herald script, supplemented in this case by a chrome and red enamelled '12/50' badge, the latter also repeated on the rear wings. This duplication was dropped on later cars, 12/50s after September 1966 relying on the boot badge only for identification.

51

RIGHT
When fitted with the optional overdrive, 1600 cars were equipped with this additional badge. It was also used on all other Vitesse models to which overdrive was fitted as an aftermarket accessory.

ABOVE
The 13/60 models carried the make and model on two badges within the number-plate recess, but no lettering on the top of the boot.

RIGHT
The Vitesse 6 was normally badged in the manner shown left, but some cars carried this additional underline (right).

ABOVE
The 2-litre cars used the script from the 1600 model, supplemented by the same 2-litre badge used on the bonnet. This car also has the overdrive badge used when factory fitted.

RIGHT
Vitesse MkIIs used the same Triumph badge as fitted to 13/60s, supplemented by this overdrive badge where applicable. Note the Vitesse badge now added to the reversing lamp shroud.

EXTERIOR TRIM

LEFT & ABOVE
Early saloons were fitted with these badges on the rear roof pillars on the Herald (left) and the Vitesse (above).

RIGHT
From late 1962, all saloons used this Triumph badge; all estates and Couriers were fitted with the badge shown in the lower picture, because of the narrower pillar.

The 948 Heralds had no supplementary boot badging, except the earliest coupés which declared their body style in script. These early cars also displayed the Herald name in gothic script on the rear wings, but from November 1959 (Y4604) both these features were deleted, although the rear wings now displayed a crossed flags logo which showed the Naval colours for 'S' and 'V' – presumably reflecting the car's Standard-Vignale origins. This crossed flags badge was fitted to all subsequent coupés and all Herald convertibles until early 1971, when it was deleted from the last batch of 13/60s.

Saloons had an additional badge on the rear roof pillars to cover a panel join: 948 Heralds and 1200s to October 1962 (GA90271) bore the legend 'Herald', the very first Vitesses (up to HB2856) carried their model name, and all other saloons had the name 'Triumph'. Estates and Couriers also had a badge in this position, but this was just a gothic 'H' – the only badge on a Courier to show its Herald origins. This badge was a simple chromed casting, whereas the saloon badges all featured enamelled panels.

'Herald', 'Vitesse', and the majority of 'Triumph' badges were blue, but 13/60s and Vitesse MkIIs were fitted with badges enamelled in black. The 12/50 was unique in having a red 'Triumph' badge.

53

CHAPTER FIVE

EXTERIOR FITTINGS

RIGHT
The 948 coupés, and subsequently 948 convertibles, were fitted with this wide, bright windscreen trim, which actually served to expand the rubber section. A narrower bright trim, with no practical purpose, came with the 1200. Early 1200s used a two-piece alloy trim, but from 1963 a one-piece flexible plastic trim was fitted.

ABOVE
The plain rubber windscreen seal used on all 948 saloons, as well as the S and the Courier.

RIGHT
Plain rubber sections used between roof and rear body on the 948 Herald saloon (left) were supplemented by bright trim on later models (right).

LEFT
Bright trim on the rear window was only fitted to coupés – late 1200s used the smaller trim as fitted to the front screen. Note that no bright trim was used on the body rubber seal.

54

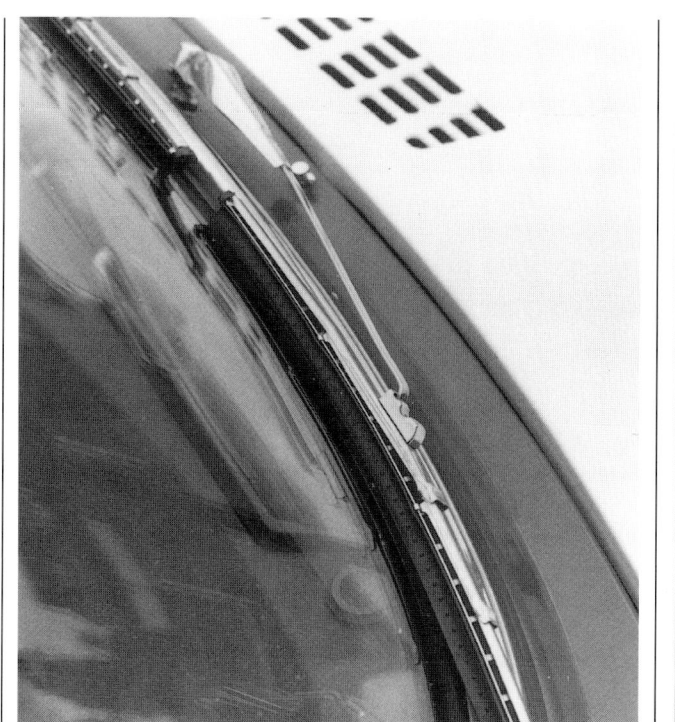

EXTERIOR FITTINGS

ABOVE
Early and late versions of the windscreen wiper arm and blade. Note, also, the two versions of washer jet. The domed version on the later car (right) was manufactured by Trico; the earlier car has Wipac units. Either version can be found on all models.

RIGHT
The pre-focus headlamp fitted to pre-1963 Heralds (top) and the sealed-beam unit used on later Heralds (bottom).

Front and rear windscreens on the original Herald saloon were retained by plain rubber sections, while coupés, and subsequently convertibles, used a different rubber section requiring a wide stainless trim to expand the rubber in order to hold the glass.

The introduction of the 1200 saw a new rubber section, using a thin alloy infill decoration, fitted to the front screen across the range. This rubber continued through all subsequent versions of the Herald and Vitesse except for the S and Courier, which kept the plain rubber of the original saloon. For 1963, the alloy decoration was replaced by a 'chromed' plastic section. Being flexible, this could be supplied as a one-piece length, having a single joint at the centre of the bottom of the screen, whereas earlier trims were supplied in two pieces with joint finishers at top and bottom of the screen. A similar decoration was also added to the base of the rear side windows and across the base of the roof beneath the rear screen on saloons, again excepting the S. The 1200 coupé was the only model to use the decoration on the rear screen itself: early 1200s retained the wide stainless trim of the 948, 1962 cars were fitted with the two-piece alloy infill to match the front screen, while later cars used the plastic version, again to match the front.

All models, including the S and Courier,

RIGHT
The twin sealed-beam units fitted to all Vitesses. Both lamps operate on main beam, the inner lamp being extinguished on dipped.

RIGHT
The side lamp fitted on most cars used this glass lens, with a supplementary orange glass dome over the flasher bulb. The lens is symmetrical, although the base unit is banded.

ABOVE
The glass lens fitted to late Heralds and Vitesses was a flatter, asymmetrical shape. The base unit was basically the same as that used on earlier cars, but a metal division between the two bulbs was no longer fitted.

LEFT
Early and late rear lamp lenses – the earlier version (left) is more prone to fading and may provide problems at MoT time.

EXTERIOR FITTINGS

ABOVE
Only the very earliest Heralds were fitted with the first style of number plate illumination (left). The vast majority of 948s and 1200s used this Lucas unit, mounted onto a small plinth painted to match the colour of the boot lid (right).

LEFT & BELOW
The aluminium shroud fitted to all estate cars (left) and the lamp that was unique to the Courier (below).

were fitted with bright stainless wiper arms and blades, the only change occurring on the introduction of the Herald 1200 when the original 'spoon' ended arm was replaced by the more universal 'spade' end. Windscreen washers were normally of a domed shape, but a squarer shape was fitted on some vehicles.

All Heralds were fitted with Lucas 7in headlamps. Before May 1963 (GA114739) the contemporary pre-focus pattern with separate bulb was used, subsequent models having the new '700' sealed-beam unit. All Vitesses were fitted with four sealed-beam 5in units, the outer units having a greater diffusing lens pattern for dipped beam use.

The majority of cars were fitted with a side-lamp unit incorporating separate side and flasher bulbs under a single glass lens with a pronounced V-shape. Late 13/60s and Vitesse MkIIs used the same lamp unit with a revised, flatter, lens. However, for a period from March 1961 to late August 1962 (GA3838-GA83026), a smaller unit using a single, double-filament bulb was fitted to the Herald range.

At first glance there would appear to be no change to the rear lamp units, but in fact two quite distinct versions were used. Before February 1965 (GA172789 and HB25738), all cars were fitted with a unit incorporating metal reflectors surrounding the bulbs, and covered by plastic lens units with no diffusing pattern. Later cars used a

ABOVE
The number-plate illumination lamp fitted to the Herald S was unique to this model.

ABOVE
This silver boot panel, fixed by studs and nuts through the boot lid, was unique to the Vitesse MkII.

RIGHT
The small shroud fitted to 13/60s (top) and the estate car shroud fitted to Vitesse 1600s (bottom).

EXTERIOR FITTINGS

LEFT
The two versions of combined number-plate illumination and reversing light fitted to Vitesse 2-litres of MkI (top) and MkII (bottom) models.

revised unit without the reflectors and with a lens design incorporating vertical ribbing.

Unlike the other lamps, the rear number-plate received many changes through the years, and across the range. As explained in Chapter 3, Heralds prior to November 1959 had a recessed unit, which was replaced by a standard chrome-plated Lucas unit which served the majority of Heralds until the introduction of the 13/60. There were three significant exceptions. The Herald S used a variation of the original recessed unit, but as it was now no longer located in a depression in the boot lid, a small bright alloy shroud was fitted. The 1200 estate used a unit mounted inside the panel, with the shroud now extended to the full width of the number-plate – dramatically different from the other models in the range. The Courier used a small round lamp unique to the model.

The Vitesse 1600 used the same arrangement as on the 1200 estate, while the 2-litre was fitted with a similar shroud which was now cut away to fit round a new chrome-plated unit incorporating a central reversing lamp. MkIIs used the same arrangement, but discreet Vitesse lettering was now included on the shroud – and to make it clear that this was the new model, a full-width ribbed steel panel, painted silver, was added to the boot lid recess.

The 13/60 models used a reduced shroud over a lamp unit now located inside the boot lid, to provide a simple means of illuminating the boot itself, although the estate retained the arrangement used on the 1200 estate.

CHAPTER SIX

WHEELS

RIGHT
The two basic versions of wheel fitted to the Herald and Vitesse differed only in the slots between centre and rim. The car fitted with the early, large slot, version (top picture) also has the blind wheel nuts fitted only to 1959 cars.

RIGHT
Wheel nut variations: left 1959; centre 1960; right 1961 onwards.

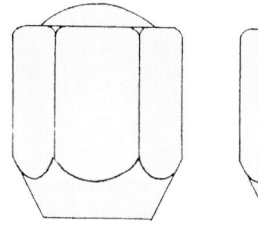

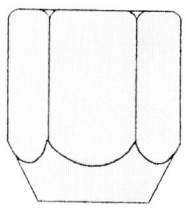

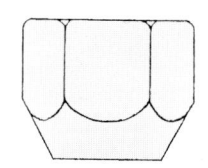

WHEELS

LEFT
The heavy duty wheel fitted to Herald 1200 estates and Couriers, and early Vitesses. 'HD' stamped in the centre panel is the only visual identification, and is normally hidden by the hubcap.

LEFT
All Heralds except the 13/60 were fitted with chrome hubcaps only – wheel trims found on earlier cars are aftermarket additions.

RIGHT
The 13/60 was the only Herald to be fitted with wheel trims. Estates used an identical design, but with a deeper recess to fit the wider wheels.

Throughout the Herald and Vitesse production run, a standard pattern of pressed steel wheel was used with only minor variations in detail. The original version featured four circumferential slots, each measuring 4½in x ⅜in, between centre and rim, and was fitted with 5.20 x 13in crossply tyres on standard 3½ D rims. Very early 948s were occasionally fitted with wheels painted in the predominant body colour, but by June 1959 the wheel colour was standardised in the usual silver-grey. Sometimes, during the 1960s, Heralds and Vitesses were fitted with black wheels; this was never a customer option, and there appears to be no rational explanation for the odd sets which occur. This early style of wheel was fitted to all 948s and continued through 1200 production until November 1964 (GA164781).

The 1200 estates were fitted with a heavy duty version of the 3½ D wheel, and these can be identified by the letters 'HD' stamped into the centre. Early Couriers and Vitesse 1600s also used this wheel, but from July 1962 (HB891) the Vitesse was fitted with a revised wheel pattern, with much narrower ⅛in slots between centre and rim, producing a significantly stronger wheel. In November 1964, this wheel was standardised across the Herald range, with a heavy duty version again being fitted to the estate and Vitesse (from HB24515).

The Herald 12/50, introduced in March

BELOW
The hubcap and trim fitted to Vitesse 1600 and 2-litre models. A similar trim, with lozenge-shaped slots, was an accessory marketed by Cosmic.

ABOVE
The Vitesse MkII 'Rostyle' trim – many have been destroyed by over-zealous fitters trying to remove the five wheel nuts. Polished finish with matt black infills was the standard presentation. Unpainted spring clips retain the trim to the wheel.

1963, was fitted as standard with the 3½ D wheel painted white and, to begin with, this was an identifying feature of the model. White sets did 'escape' on other models, however, and for a period during 1964-5 they seem to have been standard throughout the range – certainly they were never the unique feature that the early 12/50 advertisements represented them as being.

A 4½ J wheel was introduced for the Vitesse 2-litre, and also fitted to the 13/60 estate. Vitesse wheel colours were rather more stable than Herald 1200s: 1600 models were usually in silver, 2-litre models usually in white, and MkII models invariably silver.

All models except the Vitesse MkII used the same pattern of hubcap, being of chrome-plated pressed steel, with the inside surface usually finished in red oxide paint. The 1600 and 2-litre models featured a wheel trim with 12 rectangular slots, secured under the hubcap. The 13/60s were also fitted with trims, having 35 circular holes (plus an oval hole for the tyre valve), two versions being produced with different rim widths to cater for the 3½ D saloon/convertible wheels and the 4½ J estate wheels. The Vitesse MkII was fitted with a Rostyle combined hubcap and wheel trim in pressed alloy, with spring clips in the reverse to locate onto the wheel. A very similar trim was fitted to other cars in the Triumph range, but the clip dimensions were very slightly different. When fitted to the Vitesse wheel, these incorrect trims have a nasty habit of flying off: beware of trims with a ¼in diameter hole in the centre, or ones with the Triumph 'world' or BL logo in the centre. You have been warned!

Original fitment tyres on all Herald 948s, 1200s (except estate and Courier), 12/50s and 13/60s were 5.20 x 13in crossplies. The 1200 estates and Couriers used 5.60 x 13in crossplies, as did all Vitesse 1600s. Later cars had the option of 145 x 13in radials on saloons or 155 x 13in on estates and Vitesses – these became standard on Vitesse 2-litres and MkIIs.

Early Herald 948 wheels were located with ¾in deep wheel nuts, with domed, blind ends. These were replaced by open-ended nuts of similar dimension in November 1959 (G17035/Y3697), and again by ½in deep open-ended nuts with the introduction of the 1200 in 1961. This style was standard fitting for all models from this date.

CHAPTER SEVEN

MECHANICAL COMPONENTS

RIGHT & BELOW
Although Mk1 and Mk2 chassis frames differ in practically all respects, the most distinctive differences are around the rear suspension mounting. The later chassis (below) incorporated a substantial structure between the damper mountings. On assembled cars, the easiest identification is the oval hole through which passes the exhaust system on early cars (right).

It is not the intention of this book to list every variation to the mechanical parts of the cars, for it would take a whole volume to catalogue the many changes and to document the interchangability of components. Instead, this chapter will be limited to the major changes which can be identified by simple external examination. Anyone requiring more details should obtain a copy of the relevant Spares Catalogue, most versions of which are still widely available as either second-hand original publications or new re-issued volumes.

The chassis frame can be identified as one of two basic versions, with significant variations to the second one. The original, Mk1, chassis fitted to all 948s and 1200s built before May 1962 (GA80000) consisted of main rails of 3in square cross-section, with channel-section outriggers and tubular front cross-bar. The rear cross-member incorporated an oval hole through which the exhaust system passed.

The Mk2 chassis, although following the

63

LEFT & RIGHT
Another easy identification point between the chassis versions is seen on the front suspension towers. The later version incorporated an engine mounting bracket (right).

RIGHT
The front cross-bar is normally hidden by the front valance, but on a bare chassis the difference is obvious: Mk1 cars were fitted with the straight bar. The complete chassis shows the arrangement used on Mk2 Herald 1200s; 13/60s and Vitesses were identical except for the revised overrider mounting fitted to the part section.

LEFT
The brackets necessary to mount the Vitesse MkII suspension and lever arm dampers.

MECHANICAL COMPONENTS

LEFT & BELOW
The simple swing-axle arrangement used on most cars (left) and the more complex but highly effective MkII suspension (right).

same overall design, differed in almost every respect. The main rails were now 3¼in x 2¾in, with fully-boxed rear extensions and semi-boxed front outriggers; the side rails were also strengthened with fillets at the rear. The rear cross-member was given a slight upward sweep, while a reinforced damper mounting structure was added above it. The front cross-bar was given a forward sweep between the over-rider mountings, while additional bracing from the main frame was added. The main purpose of the Mk2 chassis was to cope with the additional power and weight of the Vitesse's six-cylinder engine, and so additional space was provided for the larger engine and (optional) overdrive gearbox.

In order to simplify the production line, the Mk2 chassis was also incorporated in the 1200's specification, the only differences between Herald and Vitesse versions being the mountings for the front over-rider.

The Vitesse chassis continued unchanged for the 2-litre in 1966 and was also adopted for the 13/60. In Vitesse MkII form, the chassis remained identical in all respects except for changes to accommodate the revised rear suspension. Simple flange brackets were mounted onto the main rails to support the inner ends of the wishbones, while more complex pressings were added to the sides of the original, and now redundant, damper mountings to provide the necessary location for the new dampers.

RIGHT
The two types of front lower wishbone: the early type is fitted with nylon bushes, the later uses rubber bushes.

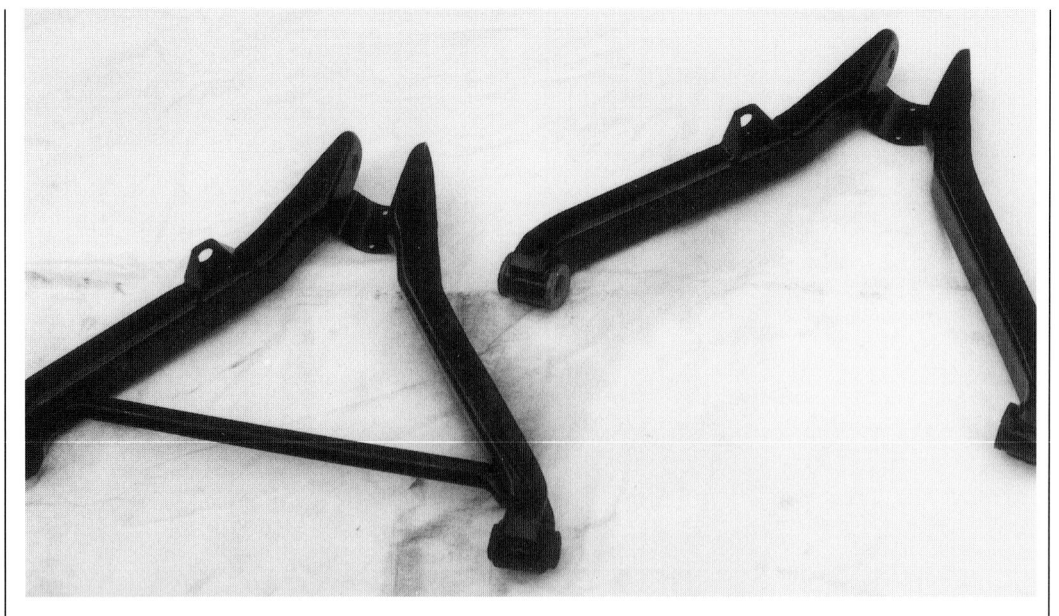

ABOVE & RIGHT
Conventional drum brakes were used on all standard 948s and 1200s: 8in drums with twin leading shoes and two external adjusters at the front (above) and 7in drums with leading and trailing shoes and a single adjuster at the rear (right).

MECHANICAL COMPONENTS

LEFT
Heralds equipped with disc brakes and Vitesse 1600s were fitted with 9in discs (left); 2-litre Vitesse models benefitted from 9¾in discs (right).

RIGHT
Alloy blocks were used to mount the steering rack on the Mk1 chassis (left); rubber mountings were introduced on the Mk2 chassis (right).

This new rear suspension was the only significant revision through the years. Various models used different strengths of rear spring, but basically all cars used the same style of leaf spring and telescopic damper. However, the MkII, while retaining the same transverse spring, now incorporated a reversed bottom wishbone and variable length drive shafts with Rotoflex couplings (generally referred to as 'rubber doughnuts'). Lever-arm dampers were mounted onto the chassis behind the drive shafts, and revised brackets and shorter radius arms were also fitted.

The front suspension, comprising combined coil spring and damper units located between upper and lower wishbones, remained unchanged between all models, again apart from different spring and damper ratings. The only minor alterations occurred in September 1959 (G8805/Y2243), when a reinforcement bar was added to the lower wishbone, and to the Courier, which did not feature the anti-roll bar fitted to all other models.

All Herald 948s and 1200s were fitted as standard with drum brakes, 8in diameter at the front and 7in diameter at the rear. The 12/50 model retained drums at the rear but gained 9in discs at the front, and this specification continued for the 13/60. The Vitesse 1600 also used 9in front discs, coupled with 8in rear drums, while the 2-litre and MkII gained 9¾in front discs. Revised front calipers were fitted to the Vitesse in August 1962 (HB11535), the same

RIGHT
These brackets were used on 1959 Heralds to locate the radiator.

RIGHT & BELOW
From late 1959, the radiator was bolted direct to the engine side valances. The full-width radiator (right) was retained until 1962, when it was replaced by the Spitfire style (below).

change occurring on special order 1200s from August 1963 (GA121841), while 1200s and 12/50s from August 1967 (GA229455/GD54349) were fitted with the larger calipers introduced on 13/60s.

The steering rack was solidly mounted on alloy blocks on the Mk1 chassis, but rubber mountings were used on Mk2 chassis. Vitesses were fitted with a lower steering ratio (requiring 3.8 turns lock to lock, compared with 3.6 on Heralds), to help compensate for the increased weight over the front wheels.

All Mk1 Heralds had a full-width radiator with a central filler cap. The radiator was initially mounted on small vertical brackets with tubular stays back to the suspension

MECHANICAL COMPONENTS

ABOVE
The remote radiator header tank fitted on early Vitesse 1600s (top) and the later radiator, introduced in 1965, fitted to all subsequent Vitesses (above).

turrets, but from November 1959 (G15548/Y3564) it was mounted directly onto the inner valances. A smaller radiator, with integral side panels and an offset radiator cap, was fitted from the introduction of the Mk2 chassis, although Couriers retained the earlier style.

Early Vitesse 1600s were fitted with a radiator with a separate, remote, header tank located between the carburettors and the rocker cover. In March 1965 (HB26150), this was replaced with a conventional radiator which also featured an overflow bottle, a system which continued onto 2-litre and MkII cars. All Vitesse radiators were mounted onto the front of the chassis, on suitable brackets located just behind the front cross-bar (these brackets appeared also on the Herald chassis but were not used).

The engine changes represent perhaps the most significant differences between the various models, varying as they do from the 34bhp four-cylinder unit in the original 948 saloon to the 104bhp six-cylinder MkII Vitesse. The original 948cc engine was the Gold Star unit already used in the Standard Pennant, and, as would be expected from an established unit, it received virtually no changes during its production. Minor internal changes entailed a new cylinder block in November 1959 (G17968/Y4673), and the new block was identified by a

ABOVE & LEFT
The single Solex carburettor fitted to the majority of 948cc Heralds (above); when twin SUs were used, this was the normal filter arrangement (left).

ABOVE
The Herald 1200 induction arrangement (top) was completely revised on the 13/60 (above).

change of colour, from the original gold (which included water pump, thermostat housing, dynamo and starter) to black (the thermostat housing was now unpainted but the rocker cover remained gold).

The single-carb unit was fitted with a Solex downdraught carburettor mounted onto a cast iron combined manifold, and equipped with a large, flat, oil-wetted gauze filter (painted black). When twin carburettors were fitted, these were side-draught SU H1 units mounted onto a cast alloy manifold, the exhaust manifold being a separate cast iron four-branch unit. The same filter was used, mounted onto a separate black-painted duct linking the two carburettors. To go with the twin carbs, a revised camshaft and high-compression cylinder head were fitted, the latter being supplied with valves located by split collets, rather than the simple collars used on the single-carb engine. The distributor cap was a Lucas unit with brown side-exit cap, the plug leads connecting to the cap on the opposite side to the engine.

Although superficially the same, the 1147cc engine introduced in the Herald 1200 differed in many details, the most obvious external identification being the fitment of a full-flow oil filter. All models were fitted with a single Solex carburettor, 12/50s having a revised manifold with a longer downpipe, and the larger bore of the twin carb 948 unit. The air filter was

MECHANICAL COMPONENTS

LEFT
The twin Solex layout on early Vitesse 1600s (top) was replaced by this Stromberg arrangement from late 1965 (bottom).

now a paper element located in a drum-shaped housing with an intake tube angled towards the radiator.

The distributor was basically the same unit as that fitted to the 948cc engine, but now in black. From September 1962 (GA86161), the cap was reversed, so that the leads now connected into the engine side of the cap. A further change occurred in February 1968 (GA238107), when the design was altered to incorporate a cap with leads connected vertically into the top.

The 1296cc engine fitted to 13/60s was a significantly different unit. The most obvious changes were to the cylinder head: de-siamesed inlet ports provided improved breathing and necessitated a reduction from 11 to 10 cylinder head studs. The engine continued to be painted black, but the rocker cover was now painted silver. The rocker cover was fitted with a radiator-style filler cap and incorporated a breather pipe to the air filter housing. The inlet manifold was now of cast alloy, and fitted with a sidedraught Stromberg carburettor, again with a paper element air filter, the housing of which was now painted silver.

From mid-1970, 13/60s were fitted with a new 1296cc engine, identified by a GK series number. This change resulted from the use of a new, larger-webbed crankshaft, involving new main and big end bearings among other components. This change is not well documented, and many high-street

71

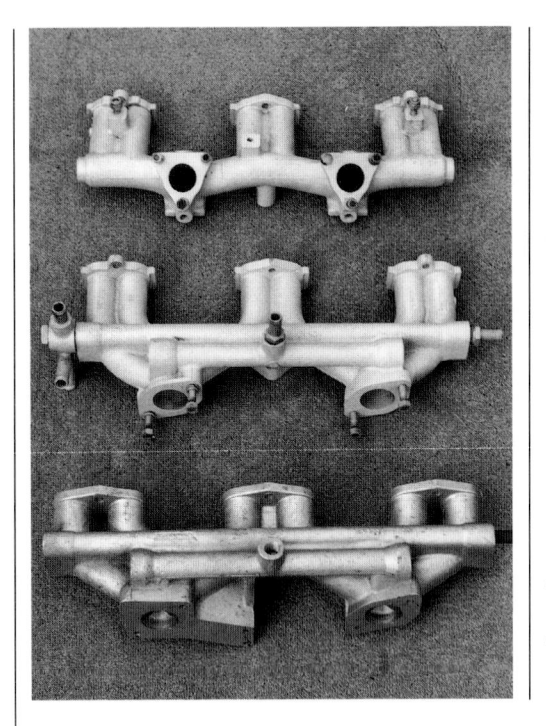

ABOVE
The five versions of fan fitted to Heralds (from left): 2-blade fitted to twin-carb 948s; 4-blade fitted to single carb 948s and early 1200s until August 1964 (GA161460E); 4-blade version with squared-off tips fitted to 1200s from mid-1964 until February 1967 (GA225581E); 4-blade 13/60 version also fitted to the last 1200s; 7-blade plastic type fitted to GK-engined 13/60s.

ABOVE
This shot demonstrates the difference between the various inlet manifolds fitted to Vitesse engines. From top to bottom: 1600, 2-litre, MkII. (John Thomason photo).

RIGHT
The six-blade Vitesse fan was replaced on the MkII by an eight-blade plastic version.

suppliers do not acknowledge the differences; if replacements are required for a GK engine, it should be noted that it is the same unit as that fitted to Toledos and Dolomite 1300s.

External changes to the Vitesse engine are less obvious than those which occurred on Heralds, and the most certain way of identifying the basic cylinder block is by the engine number, 1600s having an HB prefix, 2-litres an HC prefix (MkIIs from HC50001). All versions were painted black, including the rocker cover and air filter housing on 1600s and 2-litres (these were silver on MkIIs). However, significant changes occurred to the ancillaries, especially manifolds and carburettors. Early 1600s were fitted with twin downdraught Solex carburettors, the air filter being located beneath them. A minor change occurred in January 1963 (HB6799), when the accelerator pumps were deleted from the specification. One month later, at HB7605, the accelerator cable was changed for a mechanical linkage. From September 1965 (HB27986), twin sidedraught Strombergs were fitted, involving changes to both inlet manifold and air filter, the latter now comprising two paper elements in a single housing.

The 2-litre was fitted with an emission control valve, mounted remotely on a bracket from one of the manifold studs, the rocker cover and inlet manifold being

MECHANICAL COMPONENTS

LEFT & BELOW
The similar thermostat housing arrangement when fitted to the Vitesse: early housing (below) and later housing in full engine shot.

RIGHT
The thermostat housing fitted to Heralds before October 1965, here shown (left) with a temperature gauge sender unit. Most cars were not fitted with this gauge, so the sender unit is replaced with a threaded bolt. Later cars were fitted with a revised housing (right) with the sender unit facility now located in the water pump housing.

suitably modified. In October 1966 (HC4815), the accelerator operation was changed to a combination of levers and cable, and almost immediately changed again to cable only operation from HC4910. The MkII was fitted with a revised design of cylinder head, as used on the TR5, and yet another change to the inlet manifold. A revised camshaft and larger valves all helped to boost the performance. The new head can be identified by the fitment of integral push rod tubes – earlier heads were fitted with alloy tubes which were left unpainted. A significant internal change to the 2-litre engine was made in October 1966 (HC4501), when a revised crankshaft was fitted involving a new cylinder block, con rods, big-end and main bearings.

Single carb 948s and all 1200s were fitted with four-blade fans, but twin carb 948s had fans with only two blades. On early 948s fans were painted gold, but all later cars had black-painted fans. The 13/60s had four-blade fans with a pronounced dish to the blades, and these were painted yellow. Those 13/60s fitted with the GK-series engine were equipped with a six-blade plastic fan. Vitesse 1600s and 2-litres were fitted with fans with six alloy blades riveted to the centre boss, the blades themselves being unpainted. MkIIs were fitted with an eight-blade plastic fan.

The water pump on 948s was fitted with a thermostat housing incorporating a

RIGHT
The two styles of engine mounting used on Heralds: Mk1 cars mounted direct onto the chassis (left), Mk2 cars to the suspension turrets (right).

RIGHT
The Vitesse engine mounting arrangements.

threaded boss on its top surface. On twin carb engines this provided the location for the temperature sender unit, while on single carb engines a threaded bolt plugged the hole. Until October 1965 (GA184793), the 948 thermostat housing was fitted, but from this date a revised housing was used with an untapped boss cast on the side adjacent to the radiator hose connection. With the introduction of the 13/60, a new water pump housing was fitted, incorporating a temperature sender unit and an off-take for the water-heated inlet manifold.

The early Vitesse was not fitted with a temperature gauge, although, like the later 1200s, an untapped boss was provided on the thermostat housing. From September 1963 (HB15001), this boss was tapped and provided the location for the temperature sender unit, but in March 1965 (HB26367), a new water pump housing was introduced with the temperature sender unit now located in the main housing. This arrangement continued for all 2-litre models.

All Herald engines were supported at the front by rubber mountings located on the front engine plate. Mk1 chassis have these mountings fitted directly onto the front cross-member, behind the steering rack, but Mk2 chassis carry the engine mountings on brackets on the suspension turrets. All Vitesses also have the engine mountings on the suspension turrets, but these are now connected to the engine on brackets located on the engine block. At the rear, the engine on all models is mounted on a pair of 'bobbin' rubbers connected to the gearbox tailshaft housing. The only exception is on Vitesses fitted with the optional overdrive, in which case a single rectangular rubber is mounted onto a bracket bolted to the rear of the overdrive housing.

The gearbox, too, changed significantly over the years. Early 948s were fitted with an alloy gearbox case with integral clutch housing; this was the same unit as used in the Standard 10, Pennant and Atlas, although the tailshaft housing and rear mountings were different. In January 1960 (G29548), this was changed for a cast iron casing with separate alloy clutch housing, retaining the same internal components. The same gearbox, fitted with revised gear ratios, was fitted to the 1200, although the clutch housing was changed to a revised design in cast iron in September 1962 (GA85461). The identical gearbox continued to be fitted to all 13/60s.

The same casing – again with revised gear ratios – was also used on the Vitesse, together with a new alloy clutch housing,

74

MECHANICAL COMPONENTS

LEFT
The all-alloy gearbox fitted to early 948 Heralds; note the bracket for the adjustable clutch rod, which was deleted in mid-1960.

LEFT
When first introduced, the cast-iron gearbox retained an alloy clutch housing.

RIGHT
From late 1962 a cast-iron clutch housing was introduced. This gearbox incorporates the overdrive unit offered as a special order option on early Mk2 Herald 1200s.

75

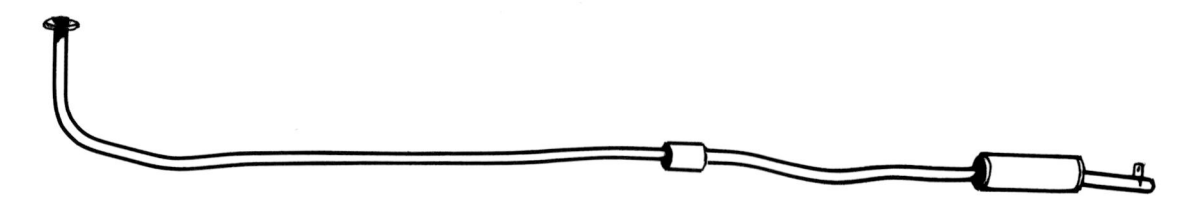

The three-part system fitted to all Heralds on the Mk 1 chassis.

The two-part system fitted to Mk 2 Heralds, 13/60s and Vitesse 1600S.

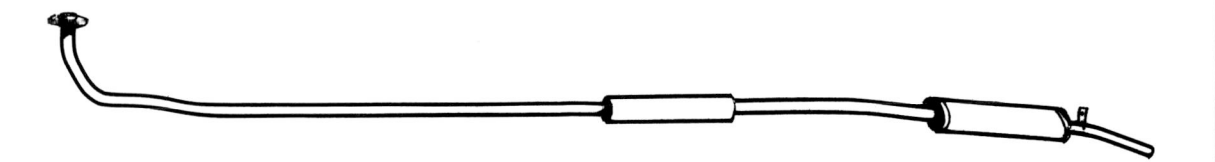

The three-part system unique to the Vitesse 2-litre.

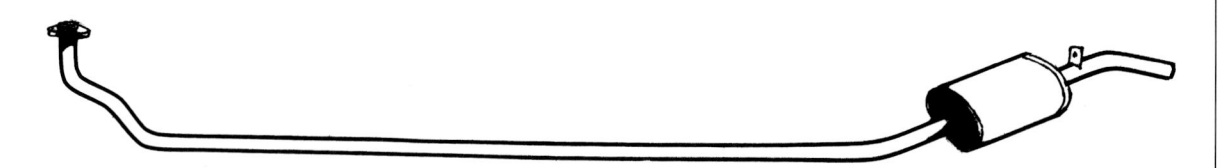

The large-bore two-part system fitted to all Vitesse Mk IIs.

ABOVE
A selection of exhaust systems fitted over the years. The front pipe differs in detail on certain Herald models from that shown.

LEFT
The alloy clutch housing fitted to all Vitesse gearboxes, showing the clutch slave cylinder position. (John Thomason photo).

with the clutch slave cylinder now mounted diametrically opposite, on the top offside. When the optional overdrive was fitted, a revised tailshaft was incorporated into the box, and the overdrive unit itself was bolted direct to the main casing to provide a complete integral unit. An electrical inhibitor switch was fitted to the top cover to restrict operation of the overdrive to the third and top gears only. The 2-litre and MkII gearboxes were fitted with synchromesh on first gear, together with an automatic switch to operate the reversing light, the switch being mounted on the top cover in a similar manner to the overdrive inhibitor switch.

A common design of differential was used on all models, although the main casing and front mounting plate were altered to match

MECHANICAL COMPONENTS

RIGHT
Drum-braked Heralds were fitted with a brake master cylinder identical to the clutch master cylinder. Note that the heater fitted to this car (left) is a Delaney Galley unit – the Smiths unit was more commonly fitted on all models. The 1959 Heralds did not have rubber covers over the master cylinder push rods (right).

LEFT
The larger brake master cylinder fitted to Vitesse models.

ABOVE
When disc brakes were fitted to the Herald, this plastic extension was fitted to the brake master cylinder. This car is fitted with the more common Smiths heater.

the revised fittings on the Mk2 chassis, and the internal components were significantly strengthened as the models increased in power. Final drive ratios also changed significantly: single carb 948s used a 4.875:1 ratio, twin carb cars using a higher 4.55:1 ratio. This was raised again to 4.11:1 for the 1200, this ratio continuing on all 1200s, 13/60s and Vitesse 1600s. All Vitesses fitted with the 2-litre engine had a 3.89:1 ratio. Herald 948s, 1200s and Vitesse 1600s were fitted with a drain plug; 13/60s and 2-litres, for some reason, were not.

Exhaust systems are not so simple. Heralds fitted with the Mk1 chassis all used a three-part system consisting of a front pipe, a small expansion box, and a tailpipe with integral silencer. The last two components were common to all models, but the front pipe altered according to the manifold and carburettor arrangements. Single carb 948s and all 1200s used a common pipe, twin carb models having a short downpipe with a larger bore. All Mk2 Heralds, including 13/60s, were fitted with a silencer/tailpipe coupled to a front pipe which differed between 1200s, 12/50s and 13/60s.

A similar two-part system was fitted to Vitesse 1600s, but with the introduction of the 2-litre a new three-part arrangement was adopted. This used a shorter front pipe and a similar tailpipe/rear silencer assembly, connected by an intermediate pipe incorporating a small diameter front silencer. The MkII reverted to a 1600-style two-part system, although the single silencer was now a larger, oval-section unit.

77

CHAPTER EIGHT

INTERIOR TRIM

RIGHT
The interior of a Herald 948. Note the thin seat squab with contrasting white piping, and the white contrasting stripe on door and side panels. This coupé has pockets on the doors, a feature unique to twin-carb 948 Heralds. Note the webbing straps under the seat cushion – these were replaced by a rubber diaphragm in 1964 (GA 141873).

BELOW
The thicker seat squabs, vertical bars on the side panels and revised stitching pattern on the seats identify this convertible as a Herald 1200. The concealed seat belt mountings were not found on pre-1965 cars.

ABOVE
The rear compartment of a 948 saloon. The squab can be folded forward to extend the boot space, the squab being retained in open and closed positions by straps and press-studs.

All the cars covered by this book were trimmed as standard in vynide, although different qualities with various textures were used through the years: a relatively smooth fabric-reinforced quality was used on 948s and early 1200s; a coarse-grained, heavier-duty quality was introduced on 1200s in 1965; and 13/60s used a softer, smoother-textured material which was also found throughout the Vitesse range.

All 948s featured white piping on the seats and a broad white stripe on door and side trims, while all other models had a single colour scheme throughout.

The 948's seat squabs were only 2½in thick, and were singularly lacking in comfort. Things improved dramatically with the 1200, the squab increasing to 3½in with softer padding to both squab and cushion; but the thinner seat continued on the S and Courier (without the white piping). Vitesse seats were improved still further with an additional ½in of padding, although following the same pattern of square top to the squab. The 13/60s and 2-litre Vitesses introduced a new style with rolled-over squab top, the Vitesse seats having more

78

INTERIOR TRIM

BELOW
Herald 1200 saloons featured a square-topped rear seat squab to match the front seats. Note the broad band in the centre of the pattern. The rear shelf, left in its bare painted state, was the continuation of the rear deck panel.

LEFT
An early Vitesse convertible, showing the door pockets which were a feature of all Vitesse models. The Herald 1200 'stitching' was retained on 1600s, but the seats were provided with additional padding. This car demonstrates the early rear side panels without seat belt mountings, and also boasts the extremely rare leather seat facings.

ABOVE
This shows the rounded top and revised stitching pattern introduced on Herald 13/60 and Vitesse 2-litre saloons. All Vitesse saloons were fitted with leathercloth covering to the rear shelf.

LEFT
The front seats on the 13/60 showing the revised squab shape, full-length stitching on the cushion, and new side trim panels.

RIGHT
A 2-litre Vitesse, showing its 13/60-style seats (but with greater padding) and the plain side trims unique to the Vitesse. (John Thomason photo).

LEFT
Herald 1200 and 13/60 estates used the same style of folding rear seat, differing only in the stitching pattern. This car can be identified as a 13/60 by the full-length stitching.

RIGHT
The enlarged recesses in the rear side panels of the 13/60 (left) and Vitesse MkII (right). Note the Vitesse's wood capping, one of many 'luxury' touches used to upgrade the six-cylinder cars.

INTERIOR TRIM

LEFT & BELOW
The dashboard pocket was supplemented by a hinged basket on 948 Heralds (below) and a map pocket on most Vitesses (left). All other cars had to rely on the tray moulded into the gearbox cover. In all cases this was left untrimmed; it was painted grey on 948 cars and Couriers, black on all other cars.

padding to give greater side support.

The rear seat on the 948 saloon was hinged to fold down to provide increased boot accommodation, but this feature was deleted on the 1200, presumably to encourage sales of the new estate. The 1200 back seat received similar upgrading in line with the front seats. The 13/60 and 2-litre Vitesse models received more fundamental changes to the rear seat, resulting in a deeper (front to back) cushion and more upright squab, designed to give more leg room to rear passengers. Convertible and estate rear passengers did not benefit from this change, as convertible space – or lack of it – and the mechanism of the estate's folding seat precluded it.

The 'stitching' pattern used on the upholstery provides another distinctive identification. The 948 models featured a series of horizontal bars at the top of the seats and across the top of the door and rear side panels (and the back panel on coupés); as previously mentioned, these were picked out in white on the side panels, but not on the seats. This style continued, in just a single colour, on the S and Courier. However, the 1200 and then the Vitesse 1600 featured a new design with a series of vertical bars on the side panels, while the seats received a plain panel across the top of the seat squab.

RIGHT
The carpet retaining clips fitted to early Heralds (left) and the alloy pressing used on all later models (right). Note the alloy treadplate which was fitted to all Vitesses, and also Herald 12/50s and 13/60s.

RIGHT & BELOW
The rubber matting fitted to Mk2 Couriers was a one-piece moulding, as seen in this view (right). Mk1 Couriers and Herald S front compartments retained the usual moulded carpet over the gearbox, with individual footwell rubbers (below).

INTERIOR TRIM

LEFT
The mottled carpet unique to the Vitesse 1600.

RIGHT
The wood cappings fitted to all Vitesse saloons: convertibles retained the door cappings, but there was no room for corresponding items in the rear compartment.

The situation changed again for the 13/60, which returned to a horizontal bar design, and for the Vitesse 2-litre, which featured a completely plain top panel.

The 948 coupés, twin-carb saloons and convertibles were fitted with elasticated pockets in the door trims. This feature was deleted from the 1200 specification, but was reintroduced on the Vitesse 1600 and remained standard on all Vitesse models. From September 1963 (HB15001), the elasticated pocket was supplemented on the Vitesse by a millboard map pocket located on the front side panel by the passenger's feet. The 948 also had supplementary stowage in the form of a drop-down basket of grey plastic-coated steel mesh, located beneath the dashboard. Hinged form the bulkhead, this was restrained in the closed position by a chrome-plated latch, and in the open position by two plastic straps which were never really strong enough to support a full basket, and have usually long since disintegrated.

Although 1200s lost this useful feature, they gained small pockets in the side panels of the rear compartments – suitable for small oddments, but otherwise of dubious value. Even these modest pockets were lost to the 13/60 rear passenger, for they were deleted to make way for larger recesses proclaimed to give extra elbow-room. These features, incidentally, were not available on the coupé, which was not

83

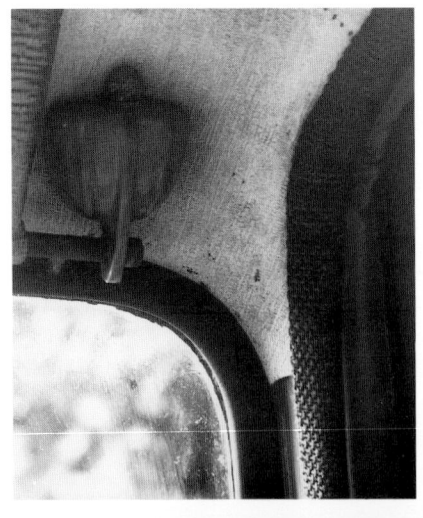

ABOVE
The roof lining on most Heralds was attached to the front edge of the roof and trapped when the roof was bolted into position (left); on Couriers and estates the trim continued over the front, and was trapped under the windscreen seal (centre); all Vitesse saloons used the normal Herald arrangements, but had additional trim across the top and sides of the windscreen frame (right).

LEFT
The roof lining on Couriers only covered the driving compartment; note the millboard trim at roof level in the rear.

intended to carry rear passengers, or on the convertible, where space was already restricted by the hood well.

Until February 1960 (G31008 and Y6403), carpets were fitted in a range of colours to match the upholstery, but after this time carpet colours were reduced to black and charcoal-grey only. From December 1960 (G59934 and Y21855), the quality of the carpet was altered, and all cars were now fitted with charcoal-grey. Prior to May 1960 the sides of the carpets were retained with flat clips by the door opening, but subsequently all models had an alloy pressing which shrouded the door seal and was retained by a series of screws through the carpet.

The carpet used in late 948s continued to be fitted to all 1200 models (with suitable alterations to accommodate the new floor pan shape resulting from the introduction of the Mk2 chassis) until December 1967 (GA236220), when the carpet quality was changed again and was now only available in black. Yet another change was made on the introduction of the 13/60, which was given a new range of four colours to match upholstery colour.

The S and Courier were fitted with rubber mats in the front compartment. Mk1 chassis cars had individual footwell mats secured by press-studs, the gearbox cover retaining the standard carpet cover. Mk2 chassis cars were given a one-piece front mat which

INTERIOR TRIM

ABOVE
The two styles of sunvisor mounting used on Heralds. The original style (left) was changed in 1960 to incorporate outer supports (right).

LEFT
Convertibles were provided with this simple outer mounting to the sunvisor.

RIGHT
Although a passenger sunvisor was not provided on Couriers, the plastic cap was still fitted in position (left). Vitesse saloons were fitted with yet another mounting variation (right).

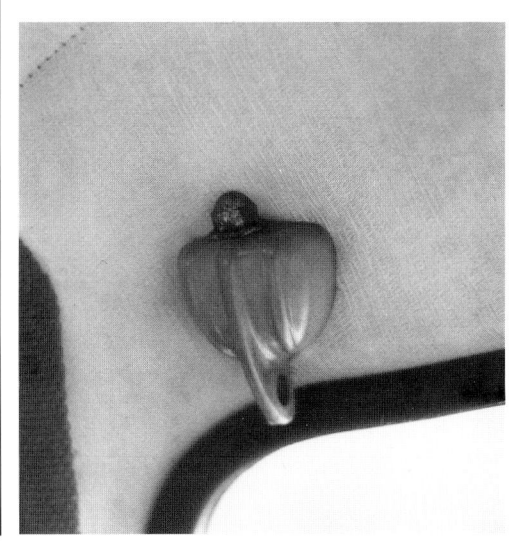

85

RIGHT
Three types of coat hook were fitted to the interior. All 948 Heralds had two metal hooks of the first pattern (left); 1200s used a second pattern, but only one was fitted to the nearside B-post. Vitesse 1600s were provided with a pair, as on the 948s. Later cars used a plastic bobbin (right); note the interior light, which was a standard feature on the Vitesse.

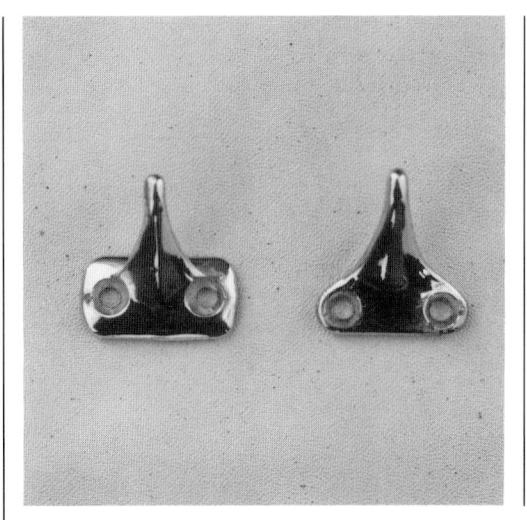

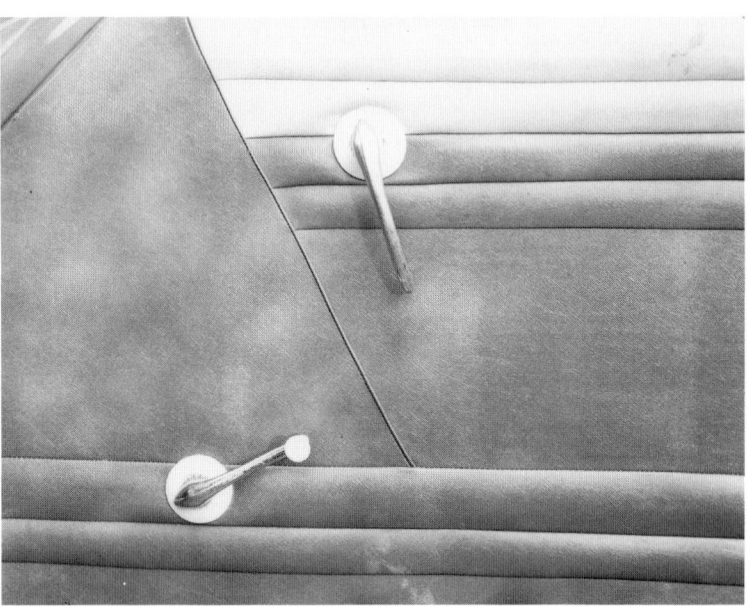

LEFT & BELOW
The interior door handle and window regulator arrangement fitted to all cars: Wisteria trim on a 948 Herald (left), black trim on most other models (below).

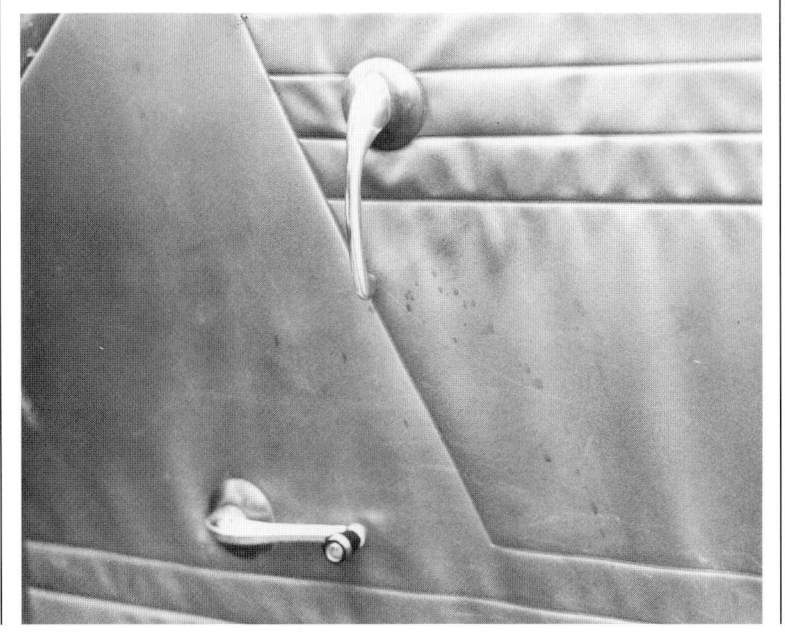

INTERIOR TRIM

RIGHT
This interior shot of an early Herald demonstrates the folding rear seat, but also shows the mounting position in the boot for the jack (BMIHT photo).

fitted over the gearbox cover and incorporated both footwells, and was again secured by press-studs. The S kept the rear compartment carpet.

All Vitesse 1600s were given a better quality carpet than that fitted to the Herald, available only in a mottled grey and black. Vitesse 2-litre carpets were available only in black, again a better quality version of the black 1200 carpet. With the introduction of the MkII, carpets matched the style and range of colours used on the 13/60.

Alloy door tread plates were fitted as standard to 12/50s, 13/60s and all Vitesses, having been a Special Order feature on 1200s (see Appendix 1). Polished wood door cappings appeared on all Vitesses (supplemented by additional rear trim in saloons), but this feature was not used on any Herald.

Roof linings did not change significantly over the years, apart from one design difference. Vitesses and Herald estates had the front edge of the lining extended across the roof bulkhead joint, and trapped by the windscreen rubber; all other Heralds had the leading edge glued to the front of the roof. The Courier's lining stopped short at the rear of the front compartment, the rear compartment having exposed metal trimmed by a 4in deep border of black millboard. All Vitesse saloons also had roof lining material covering the windscreen pillars.

Early 948s were fitted with sunvisors located only at the centre by the rear-view mirror bracket, but from June 1960 (G46844) outer supports were added, by providing grey plastic mouldings which also covered the outer roof mounting bolts. This style continued throughout Herald production, although 13/60 visors were padded. Convertibles used a different outer fixing, having a flange welded onto the end of the pivot rod and screwed onto the windscreen frame.

The Vitesse 1600 visor was similar to that fitted to the Herald, but was extended at the inner edge to shroud the rear-view mirror, while a different outer support was provided. Padded visors were introduced for the 2-litre. All previous visors were covered in the same white plastic used for the roof-lining, but the colour was changed to black for the Vitesse MkII. All passenger visors incorporated a small unframed mirror fixed to one surface; Couriers, incidentally, were not fitted with a passenger visor.

A small coat hook was provided on each side of the roof on saloons, coupés and estates; before July 1961 (GA23599), these hooks were chrome-plated metal, but later cars featured white plastic bobbins. Additionally, all Vitesses had an interior light which supplemented the dash lamp and was

RIGHT
The petrol tank found on most Heralds had a capacity of 6½ gallons (left). Note the gauge sender unit held by set screws. A larger tank (right) is found on all Vitesses and many 13/60s – this example has the sender unit retained by a locking ring, and is therefore from a 1971 car.

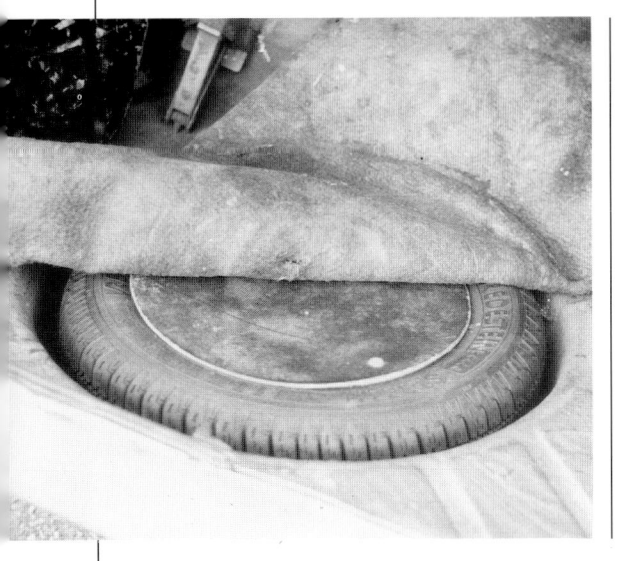

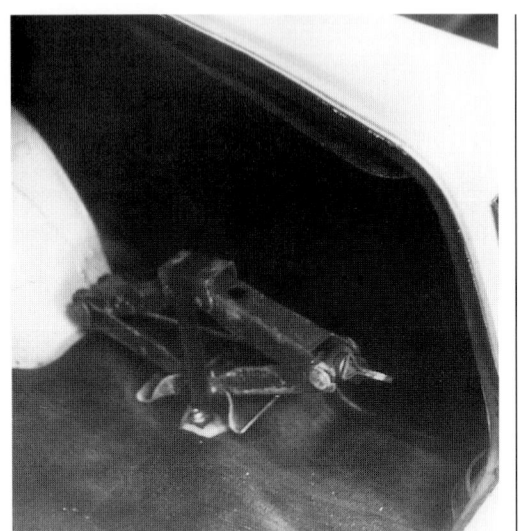

LEFT
After late 1959, the jack was moved to the rear offside wing on all except Couriers and estates.

ABOVE
All cars were equipped with a spare wheel recess in the rear floor. Most models were supplied with a hardboard disc to cover the wheel, which was held by a retaining stud.

RIGHT
Because of the restricted access on estates and Couriers, the spare wheel relied on the jack for location. This shot shows the rubbing strips provided on the floor of the estate; Couriers had plain plywood floors.

INTERIOR TRIM

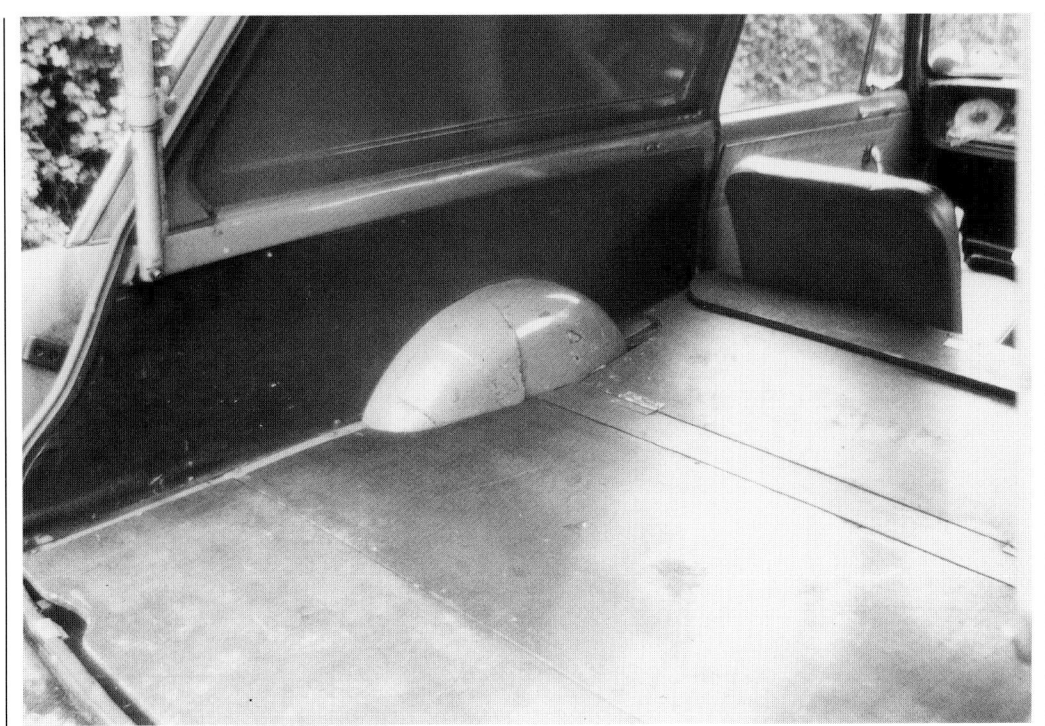

LEFT
The plywood side panels and flooring on a Courier; matt black was the standard factory finish.

RIGHT
The U-bolt mounting used on the floor to mount seat belts to pre-1962 cars (top); later cars had integral mountings located on the vertical edges of the floor (bottom).

mounted above the left-hand B-post.

Door seals fitted to 948s incorporated a fabric section which matched the upholstery colour; 1200 seals were invariably grey, while Vitesses prior to the MkII used the same seal, with black fabric. The 13/60 and MkII models used a similar seal, but again now matching the upholstery colour.

Interior door handles were common to all models, and the only change to the window winding handle was a change of knob colour – 948s used a Wisteria knob, all other cars black. The escutcheons also changed from Wisteria to black, but this occurred after the 1200's introduction in May 1961, S and Courier models again keeping the earlier colour until the end of production.

Prior to November 1967 (GA235735), all Heralds had a luggage compartment mat made of insulating felt with a black plasticised top surface. This was retained in position by press studs located across the back of the floor. All 13/60s, and 1200s from this date, were fitted with a moulded rubber mat fixed at the front corners in addition to the rear edge. All Vitesses used a mat similar to the early Herald version, but made of a thicker felt. This was supplemented by additional sound-proofing felt trapped under the mat; similar felt was also fitted beneath the rear seat cushion, under the carpets, behind the side trim panels, and attached to the roof panel above the lining.

89

BELOW
The early (left) and late (right) types of central seat-belt mounting, the change occurring between November 1964 and April 1965.

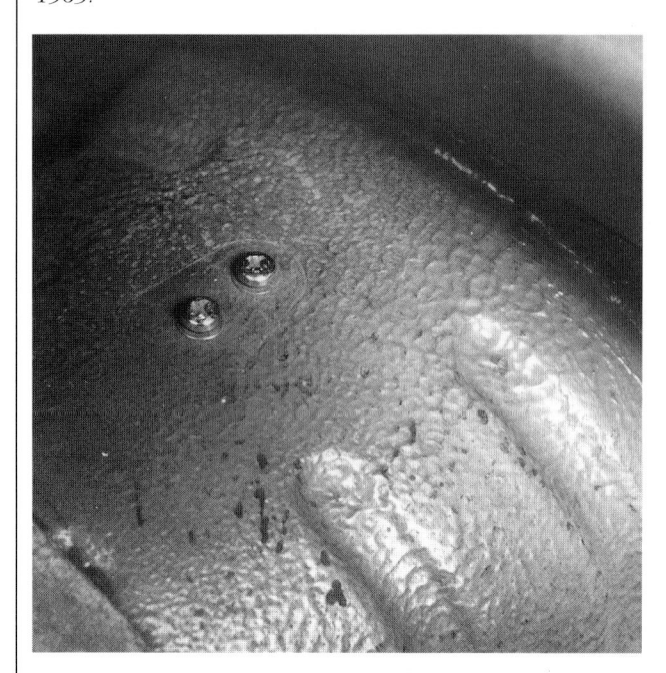

All cars, except estates and Couriers, were supplied with a circular hardboard spare wheel cover fitted with three rubber grommets.

The jack and handle on all models, again excepting estates and Couriers, were located by a webbing strap and positioned in the right-hand rear wing on most cars, but on early 948s before October 1959 (G12156) they were found in the opposite front corner of the boot, half hidden by the petrol tank. Estates and Couriers carried their jack on the spare wheel, opened-up to press against the loading floor and so locate the spare wheel.

All Herald 948s, 1200s and the earliest 13/60s were fitted with 6½ gallon fuel tanks, incorporating a small reserve facility operated by a lever located on the top of the tank. All Vitesses were fitted with a large tank of 8¾ gallon capacity, again including a reserve facility. This larger tank was fitted to most 13/60s, and the very last 13/60s and MkIIs had a further change when the tank sender unit was redesigned, the new unit being retained by a locking ring instead of six set screws. Estates and Couriers used a different arrangement, having a 9 gallon tank without a reserve located under the floor.

The estate floor consisted of sections of black-faced hardboard, reinforced with polished metal rubbing strips. The rearmost section was hinged to provide access to the under-floor spare wheel and jack. Courier floors consisted of two plywood sections for the main floor, with a third section lifting out to provide similar access to the spare wheel. Plywood side panels were also fitted in the Courier, while estates had fully trimmed side panels.

Seat belts were located at waist level behind the B-post, at floor level at the base of the B-post, and either side of the propshaft tunnel in front of the rear seat cushion on cars later than 1965. Earlier cars used a single mounting on the top of the propshaft tunnel, the change occurring in November 1964 (Vitesse HB23436) and April 1965 (Herald GA188886). Pre-1962 cars were not equipped with integral mountings, but when seat belts are fitted retrospectively this single central mounting is usually used, together with U-bolts located through the rear floor panels.

CHAPTER NINE

Dashboard & Controls

RIGHT
The original Herald saloon dashboard. The single white-faced instrument was common to all 948 saloons; the switch layout with the Herald badge above was unique to early 1959 cars.

LEFT
More comprehensive instrumentation was provided for twin-carb models, together with a metal lid to the cubby box. This coupé shows the revised switch layout introduced in July 1959.

All 948 Heralds were fitted with a dashboard made of compressed fibreboard, painted black with grey flecks (a trade finish known as Portaflec). Saloons were fitted with a single 5in instrument combining an 80mph speedometer and a small fuel gauge, and mounted directly in front of the driver. A cubby box was provided in front of the passenger, trimmed with a metal frame painted in Wisteria. Coupés, twin-carb saloons and convertibles were all equipped with a smaller, 4in, 100mph speedometer flanked by separate fuel and temperature gauges. The cubby box on these models was given a lockable hinged lid; this lid was also painted Wisteria and operated a small interior light.

All switches were in grey plastic. The heater fan was operated by a toggle switch on the extreme right-hand side of the dash, while other controls were operated by mushroom-shaped push-pull switches, arranged around a central ashtray. Initially, these consisted of a pair of heater controls

LEFT
The grey switches fitted to all 948cc models, including the S and the Courier, had their functions engraved on the face.

RIGHT
The cubby box lid was provided with a key lock (left); for 1960 a finger pull (right) was added to ease opening.

RIGHT
A Courier dashboard, showing the finger plate added behind the upper switches. Introduced in mid-1960, this feature also appears on late 948s. The horizontal bright trim was only fitted to S and Courier models.

DASHBOARD & CONTROLS

LEFT
The Herald S was usually supplied with the optional heater, in which case the dashboard was identical to the Courier's, although retaining the white-faced 948 instruments. The basic S layout used black rubber blanking plugs to cover the heater control holes.

RIGHT
The grey finish of steering wheel, column and controls can be seen on this 948 Herald. The Courier was the only 1200 model with these features.

(temperature valve and distribution) positioned either side of the ashtray, while wiper and master light switches, choke control and ignition switch were disposed in a horizontal line above. A 'Herald' script badge was located above these switches, where it received the full benefit of the interior courtesy light, which was operated automatically by both doors and manually by a lever with chrome knob integral with the bulbholder.

From July 1959, the central controls were spread into more of a semi-circle, with the Herald badge moved to below the light switch and choke control. In December 1959, the flasher warning light changed from amber to green. At the same time, the cubby box lid on coupés and twin-carb saloons was given a finger pull which partly shrouded the lock (Y3048).

In June 1960, an embossed alloy finger plate was added behind the light switch and choke control. This represented the final form of the dashboard fitted to standard 948s, and was continued with minor amendments in 'single carb' form (with single 80mph speedometer) for the Herald S; the cubby box frame and ashtray were now painted matt black, the heater controls were normally replaced by rubber grommets, and horizontal bright trim was added either side of the speedometer. When a Herald S was supplied with the optional heater, of course, the normal heater controls were fitted, but the single instrument dash was not changed for the

RIGHT
The 1960s look. From the introduction of the Herald 1200, black fittings replaced the grey items of the Herald 948.

BELOW
This new-style switch, introduced on the Herald 1200, was also used on Vitesse 1600 and 2-litre models.

ABOVE
The original Herald 1200 dashboard, showing the recessed switches. The early Vitesse 1600 featured a padded vinyl cover, but was otherwise identical.

old coupé unit on the GY twin-carb S. The S dashboard was also fitted to the Courier.

The 948's 'grey' theme was continued in the steering column, its fittings, steering wheel and column controls, which all shared a matching grey finish. Early coupés featured embossed alloy finishers on the steering wheel spokes, but these were discontinued in the summer of 1959. In January 1960 (G27426), the cover for the wiring harness to the column controls was altered to provide a clamp at its lower end. Prior to this it was only located by a felt bush, trapped by the upper column mounting.

ABOVE
The revised fascia introduced in August 1963. At this stage Herald and Vitesse dashboards were exactly the same.

DASHBOARD & CONTROLS

LEFT
These alloy trims were a feature of the Standard Pennant, but were only fitted to very early Herald coupés.

RIGHT
The Herald 1200 (left) featured a speedometer reading to 90mph (the lack of a reserve marking on the petrol gauge indicates an estate or Courier), while the early Vitesse instrument read to 110mph (right).

With the introduction of the 1200 range, the colour scheme was changed and all the fittings now became black, while the speedometer changed from white to black face. The dashboard switches now carried printed international symbols trapped into clear plastic 'windows'. The most dramatic change for the 1200, however, was the fitting of a veneered wood fascia panel, with matching ashtray and cubby box lid. Initially, the switches were still mounted directly onto the dashboard by means of recessed holes in the fascia, but from August 1963 (GA130064) a revised fascia panel was used with the switches mounted directly onto it. This change required a new ignition switch with a longer threaded shaft, but other controls continued unaltered. Just before this change, from July 1963 (GA121559), the switch for the fascia lamp was changed to a cheaper pattern – basically a flat plate bent to provide a fingergrip.

The top surface of the dashboard on early 1200s was finished with matt black paint, and small metal plates were added at the outer corners to provide extra strength from April 1961 (GA6458). From August 1963 (GA130064), a padded black

RIGHT
The superior dashboard and instruments introduced on the Vitesse in September 1963, and continued on the 2-litre.

ABOVE
The two versions of fascia lamp switch: early pattern (left), and the more common type fitted from mid-1963 (right).

vinyl cover was fitted and these cover plates were discontinued, although metal finishing strips were now fitted either side of the demister outlets. This vinyl cover had previously been standard on the 12/50 and Vitesse. Also in August, the main beam warning light (located in the speedometer, together with ignition and oil pressure lights) was changed from red to blue.

Apart from the addition of the vinyl cover, the early Vitesse dashboard was identical to its 1200 counterpart, although it now boasted a speedometer reading up to 110mph. From September 1963 (HB15001), however, the Vitesse received a major revision to the dashboard, when the single instrument was replaced by four separate units: the matching 4in speedometer and tachometer were flanked by separate fuel and temperature gauges.

The Vitesse also used similar column controls. The indicator switch, in fact, was identical to that fitted to 1200s, while the light switch featured the addition of a headlamp flashing facility, which was never offered on the 1200. The cowls around

DASHBOARD & CONTROLS

LEFT
The 13/60 received a completely revised dashboard, with recessed switches and combined wash/wipe control.

RIGHT
The three-spoke sprung steering wheel introduced on the Vitesse 2-litre.

these controls changed from metal to plastic in September 1963. When overdrive was fitted to the Vitesse, a different cowl incorporated the overdrive switch and lever behind the indicator stalk.

The 2-litre retained the same dashboard layout and fittings as the late 1600, but a TR4-style spring-spoke steering wheel, with leather rim, was now fitted, together with shorter column controls. The 13/60 also had these shorter controls, although the original Herald steering wheel was retained. The 13/60 received major changes to the dashboard, however, with 1300-style switches with T-shaped handles being fitted in a recessed central panel. The instruments themselves changed as well, a 4in 100mph speedometer being matched by a combined temperature and fuel gauge unit. The windscreen wipers and washers were now combined in a single control on the right-hand side, and the ashtray was relocated on the top surface of the dashboard.

The Vitesse MkII featured a 13/60-style dashboard, but one incorporating the usual four Vitesse instruments. Towards the end

RIGHT
The MkII Vitesse was fitted with a 13/60-style dashboard, incorporating additional instruments.

RIGHT
Late 13/60s and Vitesse MkIIs were fitted with this steering column lock and ignition switch. Note the new dashboard now fitted, without the original ignition switch.

LEFT
The grey rubber gaiters to gearlever and handbrake, grey-painted handbrake and pear-shaped gearlever knob were all featured on Herald 948s.

DASHBOARD & CONTROLS

LEFT
The black theme introduced on the Herald 1200, and featured on most models covered in this book.

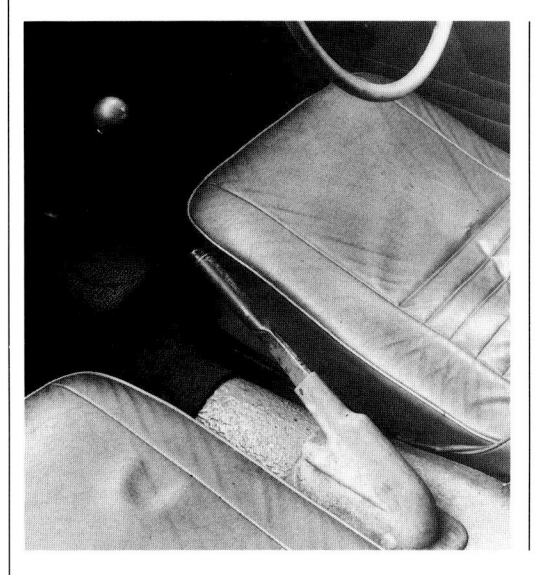

LEFT
The Vitesse knob with gear positions engraved on the top surface. One wonders why the Vitesse, which must have been bought by more experienced drivers, carried this feature, while the later Herald, which was frequently used by driving schools and novice drivers, did not.

ABOVE
The Courier was the only model with a mixture of grey and black features. Note the round grey gearlever knob, which was also fitted to the Herald S.

of production, in December 1970, both 13/60 and MkII were fitted with a combined ignition switch and steering column lock, mounted low down on the right-hand side of the steering column. This necessitated a new dashboard (without the previously dash-mounted ignition switch) which arrived at GE79339/HC57656.

Grey rubber gaiters were fitted to the gearlever and handbrake on all 948s, including the S. These were changed to black rubber on all subsequent models except for the Courier, which combined black for the gear lever with grey for the handbrake. The gearlever itself was chrome-plated on most models, the Vitesse MkII being unique in having a black plastic covering; the handbrake was painted – grey on 948, S and Courier models, and black with a black rubber handgrip on all other models.

The 948 models were supplied with a Wisteria-coloured pear-shaped gearlever knob, with the gear positions engraved in silver and black on a clear plastic disc inserted into its top. The 1200s were fitted with a plain black spherical knob – rubber until mid-1961 and hard plastic thereafter. A similar knob, in grey rubber, was used for the S and Courier. The 13/60s continued with the 1200 knob; all Vitesses used a similar item but with the gear positions engraved into the top surface.

CHAPTER TEN

OPTIONAL EXTRAS

ABOVE
The coupé rear seat, here fitted to a Herald 948. The 1200 option was identical apart from the upholstery pattern.

There are three basic categories of 'extras' which will be covered in this appendix: first, those items incorporated on the production line, referred to as 'Special Order' equipment; second, many of these items, and others, when added by the authorised dealer to a new car (or as a retro-fit to a previously registered car), when Triumph-produced or marketed with official approval, referred to as 'Accessories'; third, items falling outside these two categories will be referred to as 'Period Extras'.

From its introduction in April 1959, the Herald was always supplied in standard form with a high level of trim and equipment, so the option list was relatively small compared with its competitors. Unusually, the fresh air heater and screen washer equipment were standard fitments and the purchaser of an early Herald saloon was restricted to just four Special Order items: duo-tone paint-work, a popular choice at just £8 10s; leather upholstery at £12 15s, much less popular and therefore now an extremely rare feature; optional water temperature gauge, rather more popular, although more often added as a later fitment; and Telaflo dampers, the fourth choice, whose popularity is difficult to judge now, as it is unlikely that the original dampers will have survived over 30 years' use.

The coupé was supplied as standard in duo-tone colours, although a single colour could be specified at no additional cost, but with the disadvantage of a considerable delay in delivery. Leather upholstery and Telaflo dampers were also available, while an 'occasional' rear seat at £14 3s 4d was a surprisingly popular option, considering the minimal head room for back seat passengers. This 'occasional' seat consisted of a small rectangular cushion fitted between the rear wheelarches, and a hinged rear squab which could be folded down on top of the cushion to provide a flat luggage floor, although, unlike the saloon, this did not provide access to the boot.

By mid-1960, a laminated windscreen, white-wall tyres and 5.50 x 13 Duraband tyres (an early form of radial tyre) were added to the list of options. The convertible could be ordered with a full tonneau to supplement its standard hoodwell cover.

The Accessories list was considerably more extensive, ranging from the popular

100

OPTIONAL EXTRAS

LEFT
The veneer capping kit consisted of dashboard fascia, glove box lid, ashtray cover, two door cappings and two rear quarter cappings. Different kits were necessary for single- and twin-carb dashboard layouts, and saloon, coupé and convertible body styles (rear quarter cappings were not supplied for convertibles).

ABOVE
Cosmic bumper cappings transformed the Herald 948's austere appearance, especially in monotone finish.

Veneer Capping Kit, which provided a walnut dashboard and cappings for the doors and rear compartment, to a polished alloy exhaust tail pipe extension. White rubber bumpers were another common accessory, which proved to be so popular that they became standard on the 1200. The exterior sun visor, a feature belonging to the 1950s rather than the 1960s, was less popular. Padded interior visors, locking petrol filler cap, radiator blind and towing equipment were all items produced specifically for the Herald, while other accessories – such as fog/spot lights, reversing light, wing mirrors and rim finishers – were proprietary items given the official seal of approval.

Other items which were not publicised in official literature, and therefore are classified as Period Extras, include chrome-plated steel bumper cappings produced by Cosmic, a self-adhesive padded leathercloth dashboard top by Ritmo and a full-length folding sunroof by Webasto. Cosmic also produced aluminium door sill cappings, which anticipated Triumph's own cappings by several years.

Various tuning conversions were already developed for the Standard Pennant, and the Alexander twin-carb conversion was a favourite modification to early saloons that was effectively superseded by the factory-produced twin-carb saloon in September 1959. A more remarkable conversion was that offered by Jack Brabham, consisting of the replacement of the standard engine with the 1216cc Coventry Climax engine at a cost of £395 (over half the cost of a new car!). This was a twin-carburettor, all-alloy, overhead cam unit which transformed the car's performance to rival the then-current TR3A, with a maximum speed of around 100mph and 0-60mph in 10.8sec. A larger radiator with electric fan, larger rear wheel cylinders and Mintex M20 brake linings were included in the specification. This was a very expensive wolf in sheep's clothing, but about 20 cars, mostly coupés, were converted and there may still be one out there waiting to be discovered!

When first introduced, the 1200 continued the 948's feature of a heater as standard and the duo-tone paint was also standardised on both saloon and coupé. Both items, however, were quickly transferred to the Special Orders list as part of the economies

RIGHT
The Ritmo padded leathercloth dashboard top, like the rubber bumper accessory from Triumph themselves, anticipated later developments.

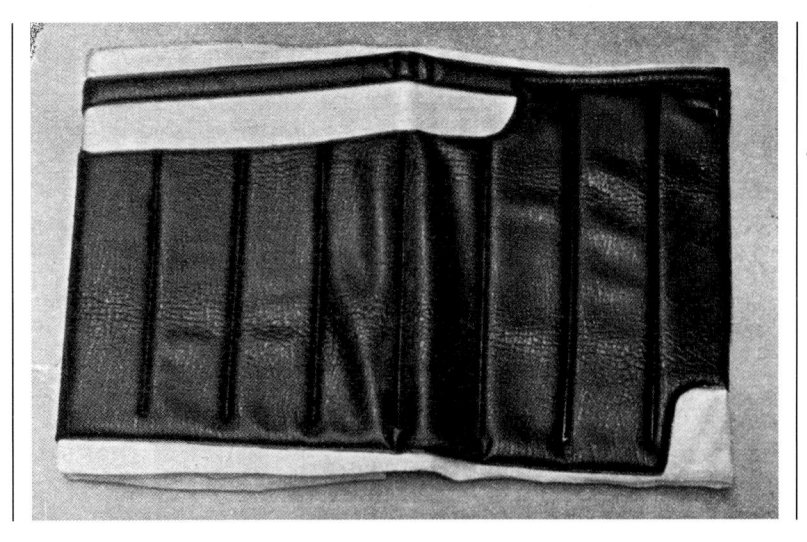

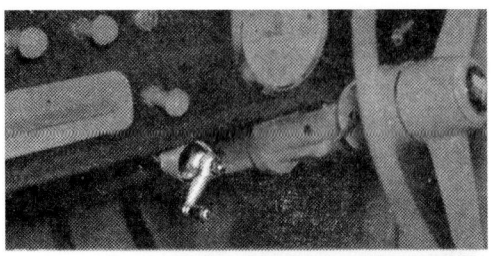

LEFT
The Imperial Venetian Blind, here fitted to the rear window of a saloon, was controlled by a remote handle mounted on the dashboard.

resulting from the Leyland takeover. It is unlikely that many cars left the production line without a heater, the new policy providing the means to increase the price of a car at the showroom, without being seen to alter its basic cost. Many saloons, and a good few coupés, did not receive the more expensive paint treatment, and over the next few years the duo-tone colours continued to lose their popularity. The coupé's rear seat continued to be available, as did the laminated screen and the convertible's full tonneau, but the Telaflo dampers were discontinued in 1963. Front wheel disc brakes became available in late 1961, and continued on all models (except the 12/50, on which they were standard) to the end of production.

Accessories available for the 1200 included many of the 948 items (such as lamps and wing mirrors) and the Witter tow-bar, while other items could be retro-fitted to the earlier cars. These included the disc brake conversion (which involved the replacement of the original stub axle and hub as well as the actual brake equipment), the bonnet lock (which provided a simple anti-theft facility to the otherwise easily-opened bonnet), and the safety harness kit (which provided a common central mounting on the top of the propshaft tunnel and a U-bolt mounted through each rear footwell). From late 1964, integral mounting points were provided for owners to fit their own choice of harness.

One Accessory available for the 1200, and not intended to be fitted to an earlier 948cc car, was the twin-carb conversion kit, which included a pair of HS1 SU carburettors, special trunking to connect to

OPTIONAL EXTRAS

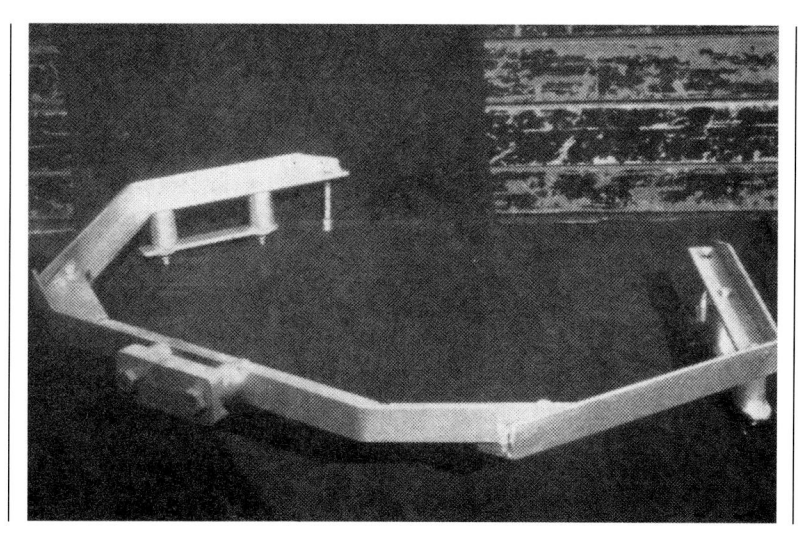

ABOVE & LEFT
A Witter tow-bar was the officially approved accessory, but this required the drilling of the bumper (above). GT Components also marketed a tow-bar, but this was mounted below the rear valance and made home-fitting a simpler exercise (left).

the standard air filter, 948-style inlet and exhaust manifolds, special camshaft and valves. It also required machining of the cylinder head to increase the compression ratio to 8.5:1, and new main bearings were also needed to protect the guarantee.

A similar kit was available for the 12/50, using twin 1¼in HS2 SUs, and requiring the cylinder head to be modified to give a 9:1 compression ratio. The 12/50 kit could, of course, be fitted to the 1200, but in addition it was necessary to change the camshaft, exhaust downpipe and distributor. These kits were readily available through dealers, and proved moderately popular.

However, further conversions existed but were not widely publicised. Only available as a Special Order on new cars, they were described as Herald 1200 and 12/50 Special Conversions. It was necessary to collect the new car direct from the Canley Works, and deliver direct to the Service Department at Allesley. Cars could only be converted if they were equipped with disc brakes and showed inter-factory mileage only.

These conversions carried the full factory warranty, and a suitable suffix was added to the Commission Number. There were five options, given the following identifications:

Conversion A Spitfire engine complete with clutch, Vitesse dashboard and instruments, twin chrome tail pipe silencer and rear axle ratio of 4.55:1. Priced at £95.

Conversion A1 As conversion A, plus Laycock D-type overdrive on third and top gears. Priced at £165.

Conversion B Spitfire engine and clutch. Priced at £65.

Conversion C 12/50 engine and clutch. Priced at £25.

Conversion D 4.55:1 rear axle ratio. Priced at £15.

All these conversions were available from 1963 until 1966. From 1964, further tuning was available in the form of the Interim, Stage I and Stage II modifications aimed specifically at the Spitfire, and obtainable through dealers or from SAH Accessories Ltd. Very few were sold, even for Spitfires, but technically they were available for the Herald, and so should be included in this section. It would have been necessary, however, to make up a special exhaust system, as a suitable unit for the Herald was not available. Wire wheels, too, were aimed at the Spitfire market, and again could be fitted to the Herald, although it should be added that the standard finish was silver paint. The chrome wire wheels offered in more recent times were never marketed when the cars were new.

The Clayton Dewandre 'Mot-a-Vac' brake servo was an official accessory, and again could be fitted to earlier cars. Girling provided the 'Powerstop' servo, which proved as popular although it was not initially the factory-approved unit. Other extras again included the Webasto sunroof, although the Weathershields version introduced on the 12/50 soon became the usual fitment.

The 1960s was the time of 'go faster goodies', and a vast range of tuning equipment was available. The most popular engine tuning establishments were Alexander Conversions, V.W.Derrington, Mangoletsi and SAH Accessories, although other well-known names offering conversions included Arden Racing, Lawrencetune, Laystall Engineering, Nerus Engineering and Speedwell Performance Conversions. Their modifications usually featured a polished and gas-flowed cylinder head, large valves, twin carburettors and special manifolds.

The Herald S continued with the same Special Options as were available on the 948 saloon, with the addition of a heater and screenwashers, which did not feature in the S's standard specification. The Courier was marketed with a similar trim specification to the S, but strangely the heater and screenwashers were listed as a combined option; the S could be ordered with either or both

ABOVE
Clayton's 'Mot-a-Vac' brake servo (left) and the equally popular 'Powerstop' from Girling (right).

of these items. A folding rear seat was available as an extra for the Courier, this being supplied by Restall's of Birmingham.

The 13/60 saw the reintroduction of the heater as standard fitment. The only Special Order item was the 12/50-style sunshine roof, which was only available on the saloon and was identified by the same RS suffix as that used on the 12/50. The estate could be fitted with a sunroof, but only as a dealer-fitted Accessory – consequently, it could be either Weathershields or Webasto, and either standard size or 'full-length'. The laminated windscreen and tow-bar were still available, but generally the Herald had now been forgotten by the accessories trade.

Aimed at a more luxurious market than the Herald, the Vitesse was more completely equipped as standard, so the Special Orders list was always very short. Laycock de Normanville overdrive, operating on third and top gear, was available throughout the

LEFT
The Webasto sunroof, most commonly found on Herald 948s, and 1200s before the 12/50's introduction.

BELOW
The useful, and popular, overdrive was operated by a third column lever, mounted in a special shroud.

ABOVE
The Restall rear seat fitted to the Courier van; Martin Walter, the Dormobile manufacturers, marketed a similar conversion, and also offered a side window kit, although the latter would render the vehicle liable to Purchase Tax.

model's production (identified by the suffix 'O' on the commission number), while the 12/50-style sunroof was available from January 1963. Leather upholstery, only available on the early 1600s, was discontinued in the summer of 1964. Duo-tone paintwork was also only available on 1600s, the Vitesse 6 convertible being the only Herald-style convertible to be available with this option. This, too, was discontinued during the production run of the 1600, cars from March 1966 being available in single colours only. The only other Special Order option was the convertible's full tonneau, and, as with the Herald, this was available throughout production.

As with the Herald, the Vitesse could be supplemented by a wide range of items from the Accessories list. Most items, such as the tow-bar, were identical to those itemised for the Herald of the relevant period, but tuning modifications were obviously unique to the model. There was no official factory-tuning equipment, although SAH Accessories had an extensive range of modifications which were given official approval, and Vic Derrington marketed a GT kit which included revised springs and dampers. However, even in standard tune the clutch was barely adequate to cope with the car's power; as no-one seems to have marketed an uprated clutch, any performance modifications were invariably accompanied by criticisms of extensive clutch slip.

The following list itemises all the Stanpart accessories that were available over the years. It is as thorough as I can make it, but probably incomplete.

OPTIONAL EXTRAS

Anti-friction throttle cable	all models
Anti-mist rear screen panel	all saloons and coupés
Badge bar	all models
Bonnet lock kit	all models
Brake kits: disc brake conversion	Herald 948 and 1200
Clayton Dewandre Mot-a-Vac	Herald 948 and 1200
Girling Powerstop	all models
Bumper kits, rubber	Herald 948 and Courier
Capping kit, wood: complete	Herald 948
door and rear quarter	Herald 1200 and 13/60
Carburettor kit, twin	Herald 1200 and 12/50
Cigarette lighter	all models
Defroster, electric	all saloons and coupés
Door buffers	all models
Fire extinguisher	all models
Fuel filter	all models
Gauges: oil pressure	1200, 12/50 and Vitesse
temperature	Herald 948, 1200 and 12/50
Heated back light	all saloons
Heater kit	Herald 1200
Lamps: fog/spot	all models
Cibié QI headlamp units	Vitesse
reverse, manual or automatic	Heralds
continental touring conversion	all models
Luggage racks, boot or roof	all models
Mats, rubber	all models
Mirrors, interior dipping and wing fittings	all models
Mudflaps	all models
Overdrive kits	Vitesse
Steering wheel: leather covered	all Heralds and Vitesse 6
wood rim	all models
leather cover	all models
Sun visor, exterior	all saloons and coupés
Tail pipe trim	all models
Tonneau covers	all convertibles
Touring kits	all models
Venetian blind	all saloons and coupés
Wire wheel kit	all models
Wheel trims: rim finishes	all Heralds
13/60-style trims	Herald 948, 1200 and 12/50
Nave-plate medallion	all models
Windscreen: laminated	all models
emergency	all models

CHAPTER ELEVEN

COMMISSION NUMBERS

In all cases throughout this book, the dates referred to relate to the actual *build date* of the relevant com-mission number. Invariably, the registration date will post-date this by several weeks, and in some cases, by several months. For instance, several changes were made to the 1200 *before* the model was announced, and it is quite conceivable that cars with a revised specification will have an earlier registration date than those with an earlier specification. Similarly, when a model was selling slowly, such as the late coupés and Couriers, or when a model was superseded, the earlier car may again have lingered in the showrooms for several months before selling. Consequently, the registration date is not an accurate means of identifying the precise age of a car, and it is always advisable to refer to the commission number wherever possible.

The British Motor Industry Heritage Trust offers a valuable service to the owner. For a small fee, it will provide the full build record for your car, including details such as original engine number, colour of bodywork and upholstery, factory-fitted options and key numbers; and in many instances, details of the distributor, selling agent and original registration number. A written request is necessary in order to monitor enquiries, and to protect the system from abuse; this should include as much information as possible (at the very least, it should quote the commission number). The details are supplied in the form of a certificate which many owners will be pleased to frame as a permanent record of their car.

I would like to record my special thanks to the BMIHT, and especially Anders Ditlev Clausager, for providing many of the details contained in this appendix, and for access to files which have helped to date many of the changes recorded in this book.

RIGHT
The official identity of a car is carried on an alloy plate mounted on the left-hand side of the bulkhead. On 1959 cars this is supplemented by an additional plate giving details of the petrol reserve tap.

TRIUMPH HERALD/VITESSE
First commission (chassis) numbers by calendar year

Year	Herald				
	948cc saloon and S (suffix SP) prefix G	948cc coupé/conv prefix Y	948cc saloon twin carb prefix GY	1200 general prefix GA	1200 export only prefix GB
1959	1 (Mar)	1 (Jan)	1 (Sep)		
1960	23752	5281	1174		
1961	60043	22127-23428 (to Jun)	10191-11392 (to Mar)	1 (Feb)	
1962	68681-69588 (to May)			45538	
1963	71462 (May)			97611	10001 (Aug)
1964	73568-73571 (BU; to Jan)			138340	14248
1965	(series continued			169447	27859
1966	for CKD kits to			200280	40728
1967	India)			222433	50588
1968				236629	56877
1969*				244523	58715
1970				248238-249873 (to May)	60490-60649 (to Apr)
1971					
1972					
1973	(to 90418; CKD India)				

Notes to table

1) Herald 948cc saloon includes 'S' model (very basic equipment, commission number suffixed SP). There seems to have been a hiatus in production between May 1962 (commission number 69588) and approx. May 1963 (commission number 71462). Intervening commission numbers were issued to CKD cars. The last commission number issued to a built-up car (BU) was 73571 in January 1964, but the series of numbers was continued until (at least) 90418 in 1973. All these later numbers were CKD kit cars for India (Standard Gazel).

2) The Herald 1200 GB-series was for export only, mostly with left-hand drive but some probably also with right-hand drive. Very many of the GB-series numbers were in fact CKD cars (assembled in Belgium and other locations). The GB series is the only series of Herald/Vitesse commission numbers that did not start with 1, but with 10001 instead. Early GB-series cars seem to have 12/50 type engines with the GD prefix, later on they revert to ordinary 1200 engines with the GA prefix, presumably once the 12/50 camshaft was standardised.

3) As the 12/50 only existed as a saloon with sunroof, all 12/50 commission numbers will have the suffix letters RS.

4) Courier vans were numbered in the ordinary Herald series but had the commission number suffix V.

5) The GG series is a real oddity. There were just 101 cars, all built in December 1967 and all destined for Puerto Rico. These cars used 1968 model year North American Spitfire type engines (with emissions control equipment), engine prefix FE. There was a subsequent Puerto Rico series, GH, 1968/69, 13/60s, again with emission controlled Spitfire engines.

6) Later 13/60s had the engine prefix GK.

7) The Vitesse 2-litre MkII started from commission number HC/50001 in 1968.

COMMISSION NUMBERS

*1969 starting numbers are approximate owing to lack of many dates on microfilmed Triumph build records.

Year	Herald			Vitesse	
	12/50 saloon with sunroof prefix GD	1200 Puerto Rico prefix GG and GH	13/60 all models prefix GE	6(1600) prefix HB	2-litre MkI/MkII prefix HC
1959					
1960					
1961					
1962	1 (Dec)			1 (Apr)	
1963	9			6731-12151 15001 (Sep)	
1964	15767			16933	
1965	29079			25171	
1966	42734			31063-34053 (to Sep)	1 (Sep)
1967	51372-55689 (to Aug)	GG 1-101 (all in Dec)	1 (Aug)		2127
1968		GH 1-409	8689		9077-12079 50001 (Aug)
1969*			37944		51905
1970			59533		55635
1971			79518-83433 (to May)		57640-58109 (to May)
1972					
1973					

8) Triumph were occasionally in the habit of 'jumping' the commission number series to the next convenient round figure plus 1. This happened either to indicate some change to the design or specification, or just to mark the beginning of a new model year (this especially from approx. 1969 onwards). Two such jumps have been recorded in the Vitesse commission number series, from HB/12151 to HB/15001 (at the start of the '1964 model'), and from HC/12079 to HC/50001 in 1968 to mark the start of the 2-litre MkII model. There may well have been other such jumps in the various commission number series but none have been traced so far.

9) It will be found that engine numbers mostly have the same prefixes as the commission numbers, with some exceptions noted above. Engine numbers are usually suffixed 'E' (for engine), 'HE' (for high-compression engine) and very rarely 'LE' (for low-compression engine).

10) Body numbers for the Herald and Vitesse models have a two- or three-letter code, often as a suffix, sometimes as a prefix. Each of these codes is unique to a particular type of body for a particular model.

11) Commission numbers are usually suffixed to indicate whether left-hand drive or an overdrive (Vitesse) was fitted. There is also normally a suffix to indicate the type of body. The following is a list of the more common commission number suffixes: DL, Saloon; RS, Saloon with sunroof; CP, Coupé; CV, Convertible; SC, Estate: V, Van; L, Left-hand drive; O, Overdrive. A Triumph Vitesse left-hand drive convertible with overdrive will, for example, have a commission number such as HC/......-LCVO.

12) The commission number prefixes on GB-series CKD cars assembled abroad have a number code indicating where they were assembled. The most commonly found is the number 1 which indicates Belgium (such as 1GB/......-LDL.

109

PRODUCTION FIGURES BY MODEL

HERALD

948			
	Saloon	1959-73*	86,129
	Saloon S	1961-63	6,577
	Coupé	1959-61	15,157
	Convertible	1960-61	8,258
All Herald 948 models			**116,121**

1200			
	Saloon	1961-70	201,143
	Coupé	1961-64	5,312
	Convertible	1961-68	43,299
	Estate	1961-68	39,821
	Saloon 12/50	1963-68	54,807
	Courier	1962-64	5,136
All Herald 1200 models			**349,518**

13/60			
	Saloon	1967-71	49,443
	Convertible	1967-71	16,091
	Estate	1967-71	17,118
All Herald 13/60 models			**82,652**

| **Grand total all Heralds** | | | **548,291** |

* CKD only for India, 1962-73

VITESSE

1600			
	Saloon	1962-66	22,818
	Coupé	1963	1
	Convertible	1962-66	8,459
All Vitesse 1600 models			**31,278**

2-litre			
	Saloon	1966-71	12,978
	Convertible	1966-71	6,974
All Vitesse 2-litre models			**19,952**

| **Grand total all Vitesses** | | | **51,230** |

These are the official production figures, which unfortunately do not separate the GY-series twin-carb 948 saloons from the G-series single-carb saloons, nor do they distinguish the Vitesse 2-litre and MkII production in 1968. However, by extrapolating the figures available, the likely splits can be determined, as shown in the table on the facing page.

Top: The early-style commission plate fitted to 1959 cars. Above centre: Early in 1960 this revised plate was introduced. Note the body style is now identified, in addition to the individual car's number. Below centre: The same plate continued to be fitted to 1200s, the suffix providing additional body style identification. Bottom: The Courier was numbered within the Herald 1200 series, but was fitted with this unique plate.

COMMISSION NUMBERS

HERALD

948	Saloon	G-series	57,977
		GY-series	10,568
		CKD	17,584
Total			**86,129**

VITESSE

2-litre	Saloon		7,745
	Convertible		4,193
Total			**11,938**

MkII	Saloon		5,212
	Convertible		2,802
Total			**8,014**

It must be emphasised, however, that these are *not* official figures, and there may be a small variance in fact, especially with the split of the G-series and CKD figures. Vitesse 1600 production officially ceased in September 1966, but 5 saloons and 11 convertibles are listed in 1967. For convenience, however, these are included in the 1966 figures.

LEFT
Top: Early Vitesse 1600s had a unique commission plate. (John Thomason photo). Centre: In 1965 a common commission plate was introduced, and included provision for paint and trim code, providing a simple means of identifying the original colour scheme. Bottom: British Leyland ownership was finally acknowledged in 1968, when this plate was introduced.

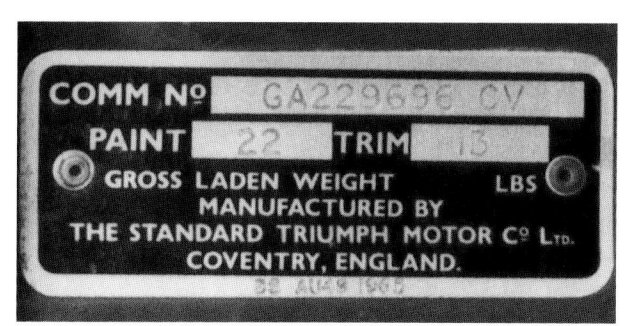

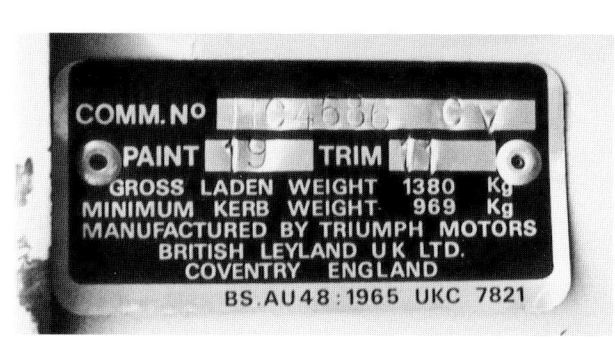

ABOVE
Engine numbers are stamped on the top face of the cylinder block, below the rearmost sparking plug. On factory-reconditioned units, this plate was riveted in position, and gave details of cylinder bore and crankshaft journal diameters.